Technology and Social Change in Belgic Gaul: Copper Working at the Titelberg, Luxembourg, 125 B.C.–A.D. 300

MASCA Research Papers in Science and Archaeology

Series Editor,
Kathleen Ryan

MASCA Research Papers
in Science and Archaeology

Volume 13, 1996

Technology and Social Change in Belgic Gaul:
Copper Working at the Titelberg, Luxembourg,
125 B.C.–A.D. 300

Elizabeth G. Hamilton

Museum Applied Science Center for Archaeology
University of Pennsylvania Museum of Archaeology and Anthropology
1996

Published by
Museum Applied Science Center for Archaeology (MASCA)
University of Pennsylvania Museum of Archaeology and Anthropology
33rd and Spruce Streets, Philadelphia, PA 19104-6324

Copyright 1996 MASCA

ISSN 1048-5325

Printed by
Cushing-Malloy, Inc.
Ann Arbor, Michigan

Cover:
Brass fibula (#338-78), faintly "silvered,"
from the Titelberg, Luxembourg, dating to
the first half of the first century B.C.
Drawing by Ardeth Anderson Abrams.

TABLE OF CONTENTS

LIST OF FIGURES vii

LIST OF TABLES viii

ACKNOWLEDGMENTS ix

CHAPTER 1:	Introduction 1
CHAPTER 2:	The Anthropology of Technology 3
CHAPTER 3:	Archaeometallurgy 11
CHAPTER 4:	The Archaeological and Historical Background: Gaul and the Roman Empire 23
CHAPTER 5:	The Titelberg and the Artifacts from the Missouri Excavations 33
CHAPTER 6:	Analytical Methodology 41
CHAPTER 7:	Results: Metal Changes at the Titelberg 43
CHAPTER 8:	Conclusions: Technology and Society at the Titelberg and in Gaul 59

REFERENCES CITED 67

APPENDIX: COMPOSITIONAL DATA 77

LIST OF FIGURES

page

Figure 1	Dendritic structure in a copper alloy (8% tin)	15
Figure 2	Fibula from the Titelberg, worked and annealed 25% zinc brass	16
Figure 3	Microstructures in copper-base metal	17
Figure 4	Pre-Roman Iron Age central and western Europe, with selected tribes	24
Figure 5	Gaul in the Augustan Roman Empire	30
Figure 6	Geological profile of the Titelberg	33
Figure 7	Plan of the Titelberg	34
Figure 8	Plan of the excavations carried out by the Luxembourg State Museum and the University of Missouri in 1968–1985	34
Figure 9	Plan of the Missouri excavation of stone foundations of workshop, associated smelters, and a side street	35
Figure 10	Stratigraphy of the Titelberg workshop(s)	35
Figure 11	Distribution of copper-base artifacts in Period 1 (second century B.C.)	37
Figure 12	Distribution of copper-base artifacts in Period 2 (100–50 B.C.)	37
Figure 13	Distribution of copper-base artifacts in Period 3 (50–1 B.C.)	38
Figure 14	Distribution of copper-base artifacts in Period 4 (A.D. 1–70)	38
Figure 15	Distribution of copper-base artifacts in Period 5 (A.D. 70–300)	39
Figure 16	Selected fibulae from the Titelberg	44
Figure 17	Scattergrams, zinc vs. tin, for Titelberg fibulae, and pins and pin shafts	45
Figure 18	A pin and a shaft from the Titelberg	45
Figure 19	Some fittings from the Titelberg	46
Figure 20	Some tacks/rivets from the Titelberg	46
Figure 21	Some tools (?) from the Titelberg	47
Figure 22	Percentages of iron vs. period for all alloys	48
Figure 23	Percentages of iron vs. period in bronze	48
Figure 24	Percentages of iron vs. period in brass	48
Figure 25	Percentages of arsenic vs. period for all alloys	48
Figure 26	Percentages of arsenic vs. period in brass	49
Figure 27	Percentages of silver vs. period for all alloys, without outlier	49
Figure 28	Percentages of silver vs. period in brass, without outlier	50
Figure 29	Percentages of antimony vs. period for all alloys, without outlier	50
Figure 30	Percentages of antimony vs. period in brass	50
Figure 31	Percentages of nickel vs. period for all alloys	51
Figure 32	Percentages of nickel vs. period in brass	51
Figure 33	Photomicrograph of Titelberg #125-82, Group 1	52
Figure 33	Photomicrograph of Titelberg #525-76, Group 1	52
Figure 34	Photomicrograph of Titelberg #133-73, Group 2	53
Figure 35	Photomicrograph of Titelberg #446-76, Group 2	53
Figure 37	Photomicrograph of Titelberg #137-77, Group 3	53
Figure 38	Photomicrograph of Titelberg #455-82, Group 4	54
Figure 39	Photomicrograph of Titelberg #9-81, Group 5	54
Figure 40	Photomicrograph of Titelberg #207-72, Group 5	54
Figure 41	Photomicrograph of Titelberg #395-74, Group 5	55
Figure 42	Photomicrograph of Titelberg #226-74, Group 6	55
Figure 43	Photomicrograph of Titelberg #727-78a, Group 7	55
Figure 44	Photomicrograph of Titelberg #912-77, Group 8	56
Figure 45	Photomicrograph of Titelberg #289-73, Group 9	56
Figure 46	Photomicrograph of Titelberg #346-82, Group 10	56
Figure 47	Photomicrograph of Titelberg #192-78, Group 10	57
Figure 48	The three brass artifacts that dated to before the invasion of Caesar	60
Figure 49	The Treveri and their various spheres of interaction	65

LIST OF TABLES

		page
Table 1	Alloys identified at the Titelberg: names and compositions	14
Table 2	Chronological chart: Gaul and the expansion of Rome	23
Table 3	The full set of artifact lots and their chronological period	36
Table 4	Frequencies of analyzed artifacts in each period, by type	39
Table 5	Analyzed artifacts from the Missouri excavations at the Titelberg, by alloy type	43
Table 6	Categorization of Titelberg artifacts by alloy components	43
Table 7	Changes in the Titelberg alloys through time	44
Table 8	Trace element means (without outliers)	49

ACKNOWLEDGMENTS

The research in this monograph was carried out as part of my dissertation for the Department of Anthropology at the University of Pennsylvania. It was financed by National Science Foundation Dissertation Improvement Grant #DBS92-24914. The materials analyzed were excavated at the Titelberg by Dr. Ralph Rowlett of the University of Missouri-Columbia. I am grateful to Dr. Rowlett and to Dr. Jeannot Metzler of the Luxembourg State Museum for allowing me access to these artifacts.

I need to thank Dr. Bernard Wailes, who has for many years provided invaluable advice, direction, and intellectual mentoring in European archaeology. Dr. Vincent Pigott taught me what I know of archaeometallurgy, though my mistakes should not be blamed on him. He introduced me to the anthropological use of archaeometry, and his advice and assistance in the laboratory were essential.

Numerous people provided technical assistance. The compositional analysis was performed by Dr. Charles Swann of the Bartol Institute of the University of Delaware. Dr. Michael Notis and his student Dong Ming were very helpful with the SEM analyses. Dr. Stuart Fleming, the Scientific Director of the Museum Applied Science Center for Archaeology (MASCA) of the University of Pennsylvania Museum, performed a large part of the computer analysis. He also allowed me unrestricted access to the laboratory and computer facilities of MASCA. Dr. Harry Rogers also provided metallurgical assistance.

The artifact drawings and maps were made by Ardeth Anderson Abrams. Paul Zimmerman of MASCA created the charts. Last-minute graphics assistance was supplied by Lisa Armstrong and Desirée Martinez.

Elizabeth G. Hamilton
November 1996

1

INTRODUCTION

Defining the research

The process termed "Romanization" is only one example of the much more general process of culture contact and culture change. Certainly the societies of Northwest Europe, those people usually if somewhat vaguely called Celts and Germans, had for centuries maintained some relations with the Greeks and Romans of the Mediterranean, as both grave goods and Mediterranean history and art will attest. In 58–51 B.C., after the Gallic Wars of Julius Caesar, those relations became those of conqueror and conquered, and the apparent result was a Gaul that used Latin, built Roman temples, worshiped the Emperor, and considered itself to be fully part of the Roman Empire. These centuries of contact and adaptation on the part of both native peoples and the Romans themselves is what has been called "Romanization," and it is often treated as if it were a long-studied and carefully analyzed cultural process. In fact, little analysis has been conducted on the transition from native "Celtic" culture to "Roman"; indeed, these terms themselves have not been adequately defined. In recent years, it has been increasingly apparent that the political changes are neither instantly nor clearly reflected in the archaeological record. The processes of cultural interaction can now be seen to have been both complex and protracted.

The use of archaeological data to investigate the processes of interaction and culture change offers both an advantage and a disadvantage. The advantage is that cultural interaction, by definition, takes place through time; it is a diachronic process. Archaeology, whose entire reason for being is to study cultures through time, is thus well fitted in its basic assumptions to investigate changes through time; there is no assumption here of static, unchanging cultures. The disadvantage, of course, is that archaeologists deal with an inevitably incomplete material record; if they are lucky enough to have a historical record as well, that record can be even more incomplete than the archaeological one.

To the scholar interested in testing theory against the archaeological record, Europe, and especially Europe in the classical period, is a gold mine. After 150 years of looting and excavation, the archaeological data base is probably the richest in the world. The written record is relatively full as well, and virtually every syllable that has come down to us has been translated, picked at, analyzed, and commented on for up to two thousand years. By and large, authors, sources, conditions of transmission and probable biases of authorship and preservation have been established, and this information lies ready to the archaeologist's hand.

But despite this wealth of potential data, the sherds are mute, and the ancient authors do not often record what we would most like to read. Thus new approaches must be developed to study changes through time, and one of the most useful involves the study of technology. The results of technological activities form the bulk of the archaeological record. It is the contention of the relatively new field of the anthropology of technology that technology is not a neutral and largely independent variable, resulting primarily from physical factors, but an integral part of the total cultural system. Technological customs can be understood only as a result of specific if often unconscious cultural choices (Lechtman 1979); therefore, the study of technological processes—the study of how people use, create, and are changed by their material, created world—gives insight into cultural processes that cannot be obtained by other means. In this monograph I concentrate specifically on one particular technological field, that of copper-base metal artifact production, and study technological change in order to examine the process of acculturation that occurred in one area in northwest Gaul.

The site of study

The Titelberg, in the Grand Duchy of Luxembourg, was occupied by the tribe of the Treveri in the Late Iron Age and Gallo-Roman periods. It is one of the most promising sites in Europe for studying the process of Romanization, principally because it was one of the few major Iron Age sites to remain occupied from the Late Iron Age (ca. 150 B.C.) through the Conquest (58–51 B.C.) and throughout the Gallo-Roman period (until A.D. 500). Recent excavations of a metalworking area on the Titelberg have uncovered a stratified assemblage of copper-alloy artifacts and working debris dating from ca. 125 B.C. to A.D. 300. The study of this material offers a rare opportunity to evaluate a technological continuum through an unusually long span of Gallic-Roman interaction.

The research program

As an initial objective, the nature of the copper-base artifact production needed to be examined. This was done by: (1) a program of compositional analysis via PIXE (proton-induced X-ray emission spectrometry), under the auspices of MASCA (Museum Applied Science Center for Archaeology), the University of Pennsylvania Museum, in order to establish the presence and quantity of eleven major and trace elements in 120 artifacts; and (2) a program of metallography (optical inspection under a metallurgical microscope), performed on the same 120 artifacts, in order to ascertain details of manufacturing technique.

These two analytical procedures resulted in a detailed picture of changes through time in the technology of both ingot production and artifact production in the period ca. 125 B.C.–A.D. 300.

Compositional studies of Roman copper-base artifacts are not unknown, but, since most of the material analyzed comes from unprovenienced museum collections, the artifacts can be dated only by stylistic criteria. This means that it is often impossible to date an artifact more closely than a century, so to assess change through time is difficult. To my knowledge, this project has produced the first chronologically controlled series of analyzed copper-base artifacts for the Roman period in Continental Europe. It is also the first to have a well-dated series of artifact analyses for the crucial Iron Age/Gallo-Roman transition period.

In addition, it is the first study to undertake full metallographic analysis in addition to compositional analysis for such a large data set, and to attempt to discern distinctive differences in metallurgical treatment through time as well as by alloy and artifact class.

2
THE ANTHROPOLOGY OF TECHNOLOGY

> Technical processes do not just produce objects, they offer us models for thinking about other things—principally ourselves. The invention of the pump gave us new ways of thinking about the human heart. The invention of the computer has recently given us completely new ways of thinking about the brain, displacing models based upon telephone systems. (Barley 1986:103)

I present below the close and detailed analytical study of a collection of copper-base metal artifacts from a single site, dating to 125 B.C. to A.D. 500. When the thesis is so baldly stated, it sounds dry and recondite, even useless, and certainly of no interest to anyone who is not obsessed with the minutiae of technical history. As I hope will become clear in this chapter, though, the study of technology and changes in technical processes can be as revealing of culture and culture change as the more common study of stylistic changes in artifacts. Technology is no mere neutral and independent variable, resulting primarily from physical factors and biological necessity, but an integral part of the total cultural system. What people choose to produce, how they produce it, and how cultural and individual attitudes, values, world view, social organization, and environment affect and are affected by technological production has been the focus of the anthropologists of technology.

What is technology?

The term "technology" as used by anthropologists in the past generally carried with it four essential components: that technology involves the manipulation of the physical world, that it involves knowledge or mental rules for manipulating physical matter, that the person doing the manipulating requires specific motor skills as well as knowledge, and that this knowledge is held by more than one person (Schiffer and Skibo 1987:595; White 1949:364).

This general consensus seemed to relegate technology to an essential, but vaguely non-cultural role in human society. As the theoretical currents changed, however (see below), the definitions used changed as well. Lechtman and Steinberg, in their important paper, "The history of technology: an anthropological point of view," use Merrill's 1968 definition:

Technologies are the cultural traditions developed in human communities for dealing with the physical and biological environment. . . . They are important not only because they affect social life but also because they constitute a major body of cultural phenomena in their own right. (Lechtman and Steinberg 1979:136)

Pierre Lemonnier, of the recently formed "anthropology of technology school," has a very explicit definition that focuses on all the elements: he breaks technology (or technique) down into five components.

1. Matter (or the material acted upon).
2. Energy (the forces of movement and transformation).
3. Objects (tools or means of work: a hammer or a factory).
4. Gestures, which move the object. Gestures are organized into linear sequences, called operational sequences.
5. Specific knowledge: the "know-how." The specific knowledge "is the end result of all the perceived possibilities and the choices, made on an individual or social level, which have shaped that technological action" (Lemonnier 1992:5).

The broadest definition is that of Pfaffenberger, also an anthropologist of technology; his definition insists that a technology must include not only techniques and material culture but also the "social coordination of labor" and the social meaning of technological activities. Techniques, material culture, and the social coordination of labor together make up a sociotechnical system (Pfaffenberger 1992:502). Comprising as it does the physical world, knowledge, know-how, social organization, meaning and values, the technological system in Pfaffenberger's sense can ultimately be defined as the entire culture. To form a definition more useful for analytical purposes, then, it is necessary to specify that the technological system is that corpus of materials, knowledge, behaviors, and values that comprise, affect or are affected by an individual's or a society's manipulations of the physical environment.

Technology in anthropology

This monograph is founded on an understanding of the role of technology in society that results from the convergence of three intellectual traditions: Anglophonic anthropology, French investigations of society and technology, and modern historians and sociologists of technology. The major developments in each tradition will be reviewed briefly.

Anglo-American anthropology

Given the complex array of factors included under the rubric "technology," it is curious that until recently attention was paid only to one—the material—aspect: tools, matter, material productions. The subdiscipline of archaeology, indeed, began when Ciracio de' Pizzacolli decided, in 1421, that Roman monuments and other physical objects offered more information about the classical past than did texts (Rowe 1965), and until roughly 1915 archaeology was in the domain of the Artifact. Artifacts were collected first as antiquarian curiosities, then as ethnic or culture group markers and guides to culture contact. They were oddly deculturized; artifacts appeared, migrated, bred, and disappeared in isolation from any human maker or social context. These early studies, while recording much of value, were, on the whole, purely classificatory and descriptive (Speth 1992:viii). The dominant paradigm of the late nineteenth and early twentieth centuries, unilinear evolution, rested on a foundation of what would later be called "technological determinism," which maintained that the forms of a society—religion, kinship, art, government, the "superstructure," in Karl Marx's terminology—were derived entirely from the society's technology and economy.

As anthropology developed, interest shifted from artifact collection to broader topics such as kinship, religion, and social organization. Cultural/social anthropology was predominantly "idealist"; i.e., it was believed that ideology, personality, and social arrangements largely determined and directed culture change, not technology or material factors (e.g., Spicer 1952). The "technological somnambulists" (Pfaffenberger 1988) detached technology from the broader culture and maintained that social structure and ideology were entirely independent of technology and technological change; any form of religion and level of social complexity could be found at any technological level (e.g., Benedict 1948:589; Boas 1940:226–267). Archaeologists maintained of necessity an interest in the artifact, but either devoted their time to largely non-interpretive chronological and typological studies (e.g., Griffin 1946; Martin et al. 1947; Montelius 1903) or emulated their cultural colleagues and wrung their material objects for stylistic clues to past kinship systems (e.g., Deetz 1965) and group affiliation (Bordes and de Sonneville-Bordes 1970; Bordes 1972). In very few cases was there an investigation of the technology as a cultural element in its own right.

One exception to this generalization was V. Gordon Childe, who from 1925 until his death in 1957 was the foremost prehistorian of Europe. Childe's early work involved a thoroughly functionalist appraisal of material artifacts. Material culture that had practical utility—tools and weapons—would be diffused rapidly from one society to another; less functional artifacts, such as ornaments and domestic pottery, would reflect local taste and have a restricted range, and these could be used to identify specific cultures (Childe 1929:248).

By 1928, however, Childe had rejected the culture-ethnic-historical school as merely a prehistoric form of political history (Childe 1958:70; Trigger 1986), and had realized that technological determinism, for all its commonsensical appeal, did not work; bronze technology in Europe was associated with very different levels of social complexity than bronze technology in Mesopotamia (Childe 1944). To get around this problem, he began to explore broader social and economic trends, explaining that the effects of technologies depended on the particular social and economic relations of production in a society. Technological developments were still seen as independent variables arising spontaneously from the workings of human intelligence, but now as variables whose effects were filtered and altered through particular cultural patterns (Trigger 1986:4–5).

In 1935, Childe visited the Soviet Union, where he became influenced by the Marxist rejection of the view that technological change resulted solely from the operation of the untrammeled human intellect. Technological change was viewed instead as occurring within "a social context that could facilitate, hinder, or block specific developments" (Trigger 1986:6). Technological change could be studied only "in relationship to the total cultural system" (ibid.). Late in his life Childe wrote, in statements that prefigure the latest works in archaeological theory,

> I realized that the environment that affected a prehistoric society was not that reconstructed by geologists and palaeobotanists but that known or knowable by the society with its then existing material and conceptual equipment. . . . A society's scientific knowledge in turn is limited by its economic and social organization. (Childe 1958:73)

Or, as Trigger rephrases his work,

> While technological changes create contradictions that bring about social and political changes, they themselves are products of specific social contexts that influence what innovations are likely or unlikely to occur. (Trigger 1986:6–7)

Childe's pioneering formulations of the relationships between culture and technology were not picked up by contemporary archaeologists. Instead, an influential anthropological paradigm of the middle of the century was neo-evolutionism, characterized by strict materialism and technological determinism (Trigger 1989:289). As neo-evolutionist Leslie White stated, "social systems are . . . determined by technological systems, and philosophies and the arts express experience as it is defined by technology and refracted by social systems" (White 1949:391). In response to this paradigm, a number of studies documenting the effects of new technology on societies were performed. Most took the technological innovation, usually an introduction from the outside world, as a given and explored the social and cultural repercussions of the new device or procedure (e.g., White 1962). A number of anthropological schools of thought, such as cultural ecology (Steward 1955), the British economic archaeology (Clark 1952; Higgs 1972), cultural materialism (Harris 1968, 1979), and the "New Archaeology" (Binford 1962, 1965) took as their foundation this sort of materialist technological determinism.

All these theories, from White through the processualists, granted enormous, indeed determining, weight to technological factors, yet treated technology as a given, an independent variable. The relationship between technology and material culture and social organization and ideology was one-way; the environment changed, new technologies emerged (in some nonvolitional way rather analogous to the appearance of genetic mutations) and therefore social organization and ideology changed (Pfaffenberger 1988; Trigger 1989:292).

In the 1970s and 1980s, dissatisfaction with the simplistic approach of the processualists grew; it was realized, as Childe had recognized, that very different cultures could result from the same environment and technology. Moreover, the processualists' fixation on testable cross-cultural generalizations neglected the true diversity of cultures, the reality of the diffusion of cultural traits, including technology, between unlike cultures, and the "unprovable" nature of most archaeological conclusions. Hence the rise of "post-processualist" archaeology, which, in reaction to the technological and environmental determinism of cultural ecology and cultural materialism, seeks the locus of culture within human relations. In this idealist school of thought, culture is comprised of human volition, choices, and power relations, almost divorced from environmental and economic constraints. The idealists indeed assign vast importance to material culture (Hodder 1982, 1989; Pocius 1991; Shanks and Tilley 1987a, b; Tilley 1990), but they are interested in artifacts, particularly stylistic variations in artifacts, purely as encoded texts, a means of communicating statements about power, ethnicity, and ideology. They do not explore the technology—the process of making—itself, but treat the artifact as a finished entity, independent of the processes and relations required to produce it. This is also true of those cultural anthropologists who have recently, after long neglect (Oswalt 1976), become interested in material culture studies (e.g., Appadurai 1986; Weiner and Schneider 1989).

The French tradition

Only in the last few years have English-speaking scholars become aware of the long-standing French interest in the relationship between technology and society (Lemonnier 1989a, 1990; Schlanger 1990). This French tradition goes back to Marcel Mauss and his discussions of cultural variations in gestures and body motions (1935). This was extended to investigation of the "technical act," where matter is affected and used in an action that is traditional, reasoned, conscious, and social. Mauss and the later French scholars were less concerned with finished objects than with "techniques," knowledge and actions, and he urged the collection of objects in all stages of fabrication and the study of how all tools were employed.

> The technical act should be apprehended throughout its unfolding, for in each of its moments, in each of its forms, in each of its gestures, the social nature of techniques find their expression. (Schlanger 1990:23)

Mauss largely divorced techniques from nature; indeed, in his gestural study, he perceived the body itself as a technical object and a technical means (Schlanger 1990:24).

André Leroi-Gourhan had a similar preoccupation with the technical act, though he embeds "technique" in the organism, in biology, rather than in society. But he shares with Mauss an insistence on the *process*, what he calls the *chaîne opératoire*.

> The tool loses its technical significance as soon as it is cut from the gestural context: Prehistory and Archaeology abound with technical objects whose significance was lost the moment the memory of their usages faded away. (Leroi-Gourhan 1957:65, as quoted by Schlanger 1990:20)

Leroi-Gourhan distinguished three levels of operation: (1) automatic, resulting from human biology and genetics; (2) mechanical, those chains of operations the knowledge of which was acquired by experience and education; and (3) lucid, which enables one to overcome problems in the operation and invent new *chaînes opératoires*. So Leroi-Gourhan finally embedded his technical actions in their social context, as Mauss did (Schlanger 1990:21).

Mauss's and Leroi-Gourhan's insights and interests in the technical process and its social context were continued by scholars such as Haudricourt (1962, 1988) and Gille (1978), in their studies of systemic and social representations of technology. More accessible to Anglophones is the work of Pierre Lemonnier (1986, 1989a, b, 1990, 1992), whose five-fold definition of technique and technology (see p. 3) constitutes an entire technology system. His work is strongly in the tradition of Mauss and Leroi-Gourhan: "Without the gestures which move it, without the matter on which it acts, and without the knowledge involved in its use, an artefact is as strange as a fish without water" (Lemonnier 1989b:156). For him, the anthropology of technology is the study of the relationship between society and technological systems, with the provisos that (1) the reciprocal effects of the technological system and the social system be considered, and (2) the most physical aspects of the technological system—matter and energy—be considered as well as stylistic traits. It is only by studying the exact operational sequences, the *chaînes opératoires*, that one can isolate "strategic moments," the operations that are essential and unalterable if a given result is to be achieved. "Examining the social control of these moments or strategic tasks is a simple and fertile means to bridge the gap between technical phenomena and other social phenomena" (Lemonnier 1986:155). Studying operational sequences likewise reveals technological variants, different ways of achieving the same end. These variants "often designate different social realities" (Lemonnier 1986:155), as seen in his field work among the Anga of Papua New Guinea. Barbed arrows, according to Lemonnier clearly more efficient than unbarbed arrows in bringing down game, are used by several Anga groups and known to others but not used by them, thus raising the question of social *choices* among even ostensibly culturally homogeneous groups. Lemonnier's work represents one of the most promising and interesting approaches to the study of technology in anthropology today. Though some of the components of his technological systems approach would be unrecoverable by prehistorians, a detailed technical analysis of raw materials, tools, and the remains of processes, as well as similar technological systems in other aspects of the culture and detailed ethnoarchaeological recording of operational sequences, would provide far more cultural information than the classification of a few stylistic traits.

The sociologists and historians of technology

Current approaches taken to the study of technology and society are derived from a number of distinct disciplines, which have converged in their interests and interpretations in the last few years. The most active field is the history and sociology of science and technology. Historians of technology have traditionally been concerned only with progress—or advancement and accumulation—in technological skills (Forbes 1950, 1955–64; Singer 1954–58; Tylecote 1976, 1986), but many have merged their interests with sociologists of technology in order to deal with the interactions of specific technology and society. This new approach has been called "contextualism" (Reber and Smith 1986; Staudenmaier 1985).

> Contextualist scholars study the technical characteristics and constraints of a particular technology (the 'design') in relation to the economic, social and political contexts ('ambience') in which they emerge and develop. (Reber and Smith 1986:1)

There are a number of variants of this basic approach: the social constructivist (Bijker 1987; Pinch and Bijker 1987), the actor-network (Callon 1987; Law 1987), and the technological system view of Thomas Hughes (1983, 1987), but they all agree on the "seamless web" of technology and society. Technological progress is in no way inevitable; which technological designs are accepted is not determined solely or even largely by efficiency (Dobres and Hoffman 1994:230), and cultural factors affect and are affected by technological development. The main differences between these approaches lie in whether the scholars focus primarily on the emergence of new technical systems, with the cultural factors considered mostly as constraints (Basalla 1988; Hounshell 1984; Hughes 1983), or whether the primary emphasis is on the cultural factors behind the technological innovation and how social choices affect technological change (Cowan 1983; Noble 1984; Schiffer 1992; Wallace 1972).

Influenced by the theoretical works of the sociologists of technology, Bryan Pfaffenberger has developed the idea of a "sociotechnical system," which "refers to the distinctive technological activity that stems from the linkage of techniques and material culture to the social coordination of labor" (Pfaffenberger 1992:497). As discussed above (p. 3), this definition thus expands Lemonnier's idea of technological systems to include the society-wide organization and control of joint labor. Pfaffenberger (1992:500) too insists that physical objects and culture form a seamless web, that sociotechnical activities are defined and shaped by society, and that society is shaped by sociotechnical system building.

It must be noted that these historians and sociologists of technology and science deal exclusively with historically known societies, and often with industrialized or industrializing societies. The cultural and ideological contexts of these societies are fairly well established already and do not need to be reconstructed, unlike those of prehistoric and early historic societies.

Current approaches in archaeology

From the 1950s there has been, remote from the theoretical considerations of the anthropologists of technology, a thin though steady trickle of archaeologists with physical science training, who studied artifacts with an eye to the classification of technical features. Though ceramic temper and physical characteristics of pottery had long been used as cultural markers (Orton et al. 1993; Rice 1987; Shepard 1965), it was only in the 1950s and 1960s that such features of metal artifacts as alloy composition and trace elements were used as distinctive cultural and chronological features (Butler and van der Waals 1966; Junghans et al. 1960, 1968; Liversage and Liversage 1989; Northover 1980; Pittioni 1982; Stutzinger 1984; Waterbolk and Butler 1965). Though an effort to use trace element composition as a guide to the original ore source failed in the case of copper (Tylecote 1970; Tylecote et al. 1977), it was realized that elemental composition, whether the result of deliberate alloying or not, could be used, like stylistic variation, to determine cultural groups and trace the spread of artifacts and peoples.

More recently, archaeologists have been expanding on this long-standing "technological-feature-as-cultural-marker" tradition, and using technological traits to delineate processes of culture contact. These studies document the transfer of technological traits into a new area and their subsequent transformation (Ehrenreich 1991; Hosler 1988; Wright 1985, 1986), assess evidence for craft specialization (Levy 1991; Nelson 1991; Rice 1981) and manufacturing systems (Franklin 1983; Northover 1989; Peacock 1982), or construct models for social change (Geselowitz 1988; Kingery 1986; McGovern 1989; Schiffer and Skibo 1987). Other archaeologists follow in the footsteps of Leslie White and examine production in terms of functional time, raw material, and energy constraints (see essays in Torrence 1989). In the sense that all of these studies are "contextualist" and regard technology and culture as a seamless web, they participate in the same intellectual construction as the modern sociologists of technology. Their use of technological traits as cultural markers and their association of technology with economic and political changes is not particularly new in archaeology, but these studies can be very informative.

What is more innovative is the apparently independent invention of the concept of "technological style" by archaeometrical archaeologists. Hughes, a sociologist of technology, uses the phrase in his 1987 article to refer to different ways of producing the same result in different areas. For example, he contrasts the London style of building numerous small power plants with the Berlin style of just a few large plants, and explores the technical and political reasons behind the variation. His idea of style is confined to a single technological sphere.

In contrast, the archaeologists who invented the concept of "technological style" extended it to the entire technological universe of a particular society. Heather Lechtman, the principal scholar in this area, has constructed a hierarchy of investigations for an anthropology of technology (Lechtman and Steinberg 1979). First, there must be extensive laboratory analyses of both the artifacts and the processes of manufacture, to identify operations that were essential and what Sackett (1982) calls "isochrestic variation," i.e., variations resulting from technologically equivalent, though mechanically different, operations. Next, one should examine the development of a single technology within a single culture area, however that is defined, and show how technological changes correlate with other cultural changes. Processes of innovation and retention of technologies must be examined, as in the retention of bronzeworking techniques for the manufacture of Luristan iron swords (Pigott 1980:446), to give evidence of common approaches to materials through the whole culture area and time.

Lechtman and others maintain that a common attitude towards materials often is reflected in these various technologies, and that one can "elicit . . . from the technology information about its own symbolic message, and about cultural codes, values, standards, and rules that underlay the technological performance" (Lechtman 1977:17). She illustrates this by showing how Inca gold work and textiles both share a preoccupation with integrity or "essence." The Inca developed a special technique called depletion gilding to make low gold alloys look like pure gold without plating or gold leaf, and Inca textiles are characterized by lack of appliqué work, with the color and design an integral part of the fabric. In both these cases, the valued ingredient—the gold or the color—are inside the object, not merely a surface film. She relates these common technological traits to elements, known or putative, of the Inca value system.

Lechtman's program is ambitious, and in fact most studies carried out either by her or under her intellectual influence have confined themselves to the study of specific technologies, either in one society (Childs 1991a, b; Epstein 1993; Lechtman 1988) or cross-culturally (Steinberg 1977). The concept of "technological style" in archaeology has some problems; many scholars use a "fairly passive definition of style" (Hegmon 1992:529), and in addition, the concept is a static one, with no discussion of technological change. This may aid interpretation of what are in fact static technological situations (e.g., Epstein 1993), but if the central archaeological problem is understanding culture change, then the concept needs expansion.

Some of the most interesting work in the anthropology of technology has been done on sub-Saharan

African metalworking. Because the indigenous copper- and ironworking industries survived for so long—the last furnace in the Côte d'Ivoire that smelted for the market closed down in 1983 (Childs and Killick 1993:325)—researchers who study native metalworking are presented with, even forced to study, the entire panoply of cultural practices surrounding metal production, practices that have vanished in the rest of the world (Childs 1994). From these studies, most done quite recently, comes eloquent evidence of how blurred the boundary can be between "utilitarian" iron tools and "expressive" artifacts (Childs 1991a), as well as how complex are the social contexts that surround the adoption of new brassworking technology (Wade 1989). Childs (1991b) has also, using metallographic analysis, carefully reconstructed the "technological styles" of copper- and ironworking in southeastern Zaire.

The African studies clearly illuminate the role of ritual and gender in metal production (Childs and Killick 1993), the issues of power and control of the output (Wade 1989), the role that metal production might play in political centralization (De Barros 1988; Childs and Killick 1993; Kusimba et al. 1994), the social role of the smiths (Childs and Killick 1993; Rowlands 1971), and the social meanings of the artifacts produced (Childs 1991a; Wade 1989). It is even possible to examine how metallurgical processes were explained "by drawing upon indigenous theories of natural and social order" (Childs and Killick 1993:319) and how these beliefs affected production. As Childs (1994) notes, ethnoarchaeological and ethnohistorical studies in Africa have, without their authors quite realizing it, provided the strongest illustration yet for the "seamless web" of technology and society.

Virtually all these studies, in Africa or elsewhere, have been performed on historically or protohistorically known societies, which raises the question of whether it is possible to extend all the various conceptions of technological style, technological system, and sociotechnical system into prehistory. The post-processualists have indeed constructed elaborate interpretations of patterns of prehistoric material culture (e.g., Hodder 1984; Tilley 1984), but, as Trigger (1989:351) points out, "so far no archaeologist has discovered how to get beyond speculation in interpreting the cultural meaning of such regularities for early prehistoric times." Lechtman's concept of "technological style," searching for congruences between various technologies, may offer some evidence beyond simple speculation; it is not clear how Lemonnier's technological system—gestures, raw material, technique, and all—could ever be fully reconstructed. And even if it could, Lemonnier points out that it is not enough to demonstrate that a relationship exists between a technological system and particular features of social organization.

> Every social theory of material culture should, also, necessarily, explain the specificity of these relationships; why they exist in the given case, *and* try to understand cases when they do *not* exist. (Lemonnier 1989b:157)

Most recently, Dobres and Hoffman (1994) have synthesized the work of Lechtman, Lemonnier, and the sociologists of technology into a theoretical framework that is rooted in practice theory. Their emphasis is on the "social agency" of technologies, using a model of scale, dynamic contexts, materiality, and social theory. In keeping with post-processual stress on power relations in society, they are most interested in political (power) relations among various groups in a particular society. Power relations can be maintained and measured by access not only to means of production but also to specialized technological knowledge. Dobres and Hoffman believe that a perspective rooted in practice theory (Ortner 1984) is one that can be used to understand the dynamic social processes involved in technological production. Practice theory takes account of individual agents acting inside a social habitus but with an incomplete knowledge of social rules, incomplete knowledge of their situation, and with their actions producing unexpected consequences. Human agents are not automata, acting according to fixed social rules; they are strategic actors, but other actors, social groups, the built and natural environment, traditions, and power relations impinge on the individual's actions. They claim that practice theory, which takes account of individual agency, social relations, and consequences through time, can provide a link between individual actions and larger-scale social transformations (Dobres and Hoffman 1994:223), but as with so many theoretical presentations in archaeology, it is difficult to see how a reasonably testable and empirically compelling scenario predicated on practice theory could be constructed from archaeological evidence.

Theoretical underpinnings of the present research

Theoretical elements from all the traditions discussed above are incorporated in this monograph, though I rely particularly upon the "sociotechnical system" approach of Pfaffenberger. I consider two facets of one specific technology, that of copper-base metalworking, and attempt to connect changes or lack of changes in this technology with historically known changes in the political, economic, and social spheres of the Iron Age and Gallo-Roman world. I regard technology as a seamless part of culture. The interactions between the Gauls and the Romans, as reflected in the material culture, are by no means simple; the results of this study suggest that contact with the Mediterranean was associated with the introduction of new alloys, which had to be obtained in new ways, which were used for new and strictly Gallic, not Mediterranean, purposes, and which had specific meanings attached to them. The manufacture and distribution of metal artifacts after the Roman Conquest, it will be argued, was accompanied by the creation of a completely new sociotechnical system.

3
ARCHAEOMETALLURGY

> Behind the vaulting intellectual accomplishments of mankind in art and music, literature and philosophy, we are beginning to glimpse, through the dust and fumes and smoke of ten thousand years of mining and metalworking, the contributions to human comfort and material progress that were made by the begrimed miner and the sweating blacksmith. (Raymond 1986:xi)

The term "archaeometallurgy" was coined relatively recently, and the field it describes, along with other areas lumped together under the term "archaeometry," still has not achieved full integration in more traditional archaeological circles (Cleere 1993). As Dunnell (1993:161) comments, "many, if not most, archaeologists regard archaeometry as a sometimes interesting, largely irrelevant, and definitely optional endeavor." There are a number of reasons for this: first, archaeologists tend to lack training in the hard sciences and instrumentation, and thus lack understanding of the kinds of questions these scientists can ask and answer; second, despite the emphasis laid by such founding fathers of European archaeology as Montelius, Reinecke, and Childe on the cultural and economic importance of metals, the "underlying technological and social processes [are] only dimly understood" (Cleere 1993:176); and third, archaeometry is often expensive and, for the above reasons, is seen as an expendable frill when budgets are tight.

Despite this, much of the blame for the lack of integration between archaeometallurgy and archaeology must lie with the archaeometallurgists. Many of the scholars conducting archaeometallurgical studies have been trained as materials scientists, and their primary interest has always been the history of technological development, largely divorced from the surrounding culture (e.g., Forbes 1950; Tylecote 1976, 1986, 1987). They have not viewed archaeology as a science that asks specific, often testable, kinds of questions, but rather suggest that the value of archaeometallurgical studies to archaeologists lies in the enhancement of "the empathetic understanding and sense of wonder derived from viewing the objects" (Vandiver and Wheeler 1991:xxii), an approach that Dunnell (1993:164) disdainfully compares to "the non-scientific archaeology of the Sunday newspaper's feature section." Much of the work performed by archaeometallurgists is useless for the development of archaeological models of social, economic, and cultural relations (Ehrenreich 1991). The objectives of archaeometallurgy should not be merely to study the history of smelting, or to enhance the appreciation of artifacts, but

> to augment our understanding of the rise of craft specialization, the organization and importance of prehistoric industries, the effects of new technologies on societies, the extent and limits of cultural contacts, and the impetus and alterations required to change rudiments of societal infrastructure. (Ehrenreich 1991:55)

The justification for the detailed study of metal artifacts, remains of smelting and mining, and details of manufacture, then, is that without this information no accurate picture can be obtained of the sociotechnical system of metalworking, a sociotechnical system that comprises techniques, material objects, the social coordination of labor, and the ideological meanings of production activities (Pfaffenberger 1992).

Archaeometallurgy and anthropology

The answers to the technological and social questions raised by Ehrenreich, thus, must rest in part on the detailed examination of the behavior of the material.

> Basic to our understanding how, why and where [these artifacts] were made, how they were used or preserved, and how they came to be deposited is a broad knowledge and appreciation of the behaviour of these materials, a knowledge which is still shockingly incomplete. Only then can we see the ways in which users of these materials in the past saw material properties, which properties were developed and exploited and which were regarded as unimportant. (Northover 1989:213)

Northover (1985, 1989) has drawn up a useful scheme of interrogation for metal artifacts, a scheme that may be applied to any archaeological material (adapted from Northover 1989:215 and Scott 1990:2):

Metallurgical
> What is it made of?
> Where did the materials involved come from and in what form?
> How was it made?
> Why was it made from that material?
> Why was it made the way it was: ignorance, indifference or design?
> How was it used and how did its use affect it?
> Can the composition and technology be matched elsewhere?

Archaeological
> What is it and what was it used for?
> Where did it come from?
> How does it relate to other material on the site?
> Why was it made like that?
> How did its use affect users?
> How did it get to the site?
> Why was it found where it was?
> Why is it in its present state?
> What period does it date to?

It should be noted that although stylistic analysis *per se* is not included in the purview of archaeometallurgists, stylistic features can be very useful for dating and for suggesting cultural affiliations and available tools and techniques.

Northover, however, is not an anthropologist, so it is necessary to add a third category of questions:

Anthropological
> What appear to be the values attached to this material (as seen, for example, by kind and placement of deposition in graves)?
> How and when do the techniques of manufacture change, and how does this correlate with other technological and cultural change?
> Is there evidence for craft specialization?
> What was the social position of the craftworkers?
> Do the technology and material reflect culture contact, and if so, how?
> Is the metalworking technology congruent with other kinds of technology, and can this congruence reflect and embody societal values?

This is only a sample of the kinds of questions that should be asked of the body of data revealed by metallurgical and site analysis. In the last few years, some studies dealing with ancient metalworking have indeed incorporated these broader questions of economy and society as a necessary part of their research, thus making them truly anthropological (e.g., Alexander 1981; Childs 1991a and b; Ehrenreich 1985, 1991; Epstein 1993; Geselowitz 1988; Glumac and Todd 1991; Heskel and Lamberg-Karlovsky 1980; Hosler 1988; Lechtman 1979, 1980, 1984, 1991).

Methodology

Compositional analysis

The most commonly used analytical laboratory technique in archaeometallurgy has been compositional analysis, also called elemental analysis. This is simply the determination, by a variety of techniques (including X-ray fluorescence, atomic absorption spectroscopy, ultraviolet spectroscopy, wet chemistry, neutron activation analysis, electron microprobe, and proton-induced X-ray emission spectroscopy), of the presence and/or quantity of particular elements in the metal, such as the percentage of tin or iron or sulfur. These compositional analyses have been performed for two purposes: alloy determination and ore sourcing. Alloys are deliberate mixtures of a primary metal and other elements such as tin or carbon (or deliberate use of ores that naturally contain useful elements such as arsenic or zinc). Determining the composition of the alloy reveals the technological capacities of the smiths and the metallic properties they considered valuable.

In the 1950s and 1960s it was suggested that the micro-impurity (trace element) patterns of a metal could indicate from which ore source it came (Hartmann and Sangmeister 1972; Junghans et al. 1960, 1968; Pittioni 1982). Strong doubts were raised about the feasibility of micro-impurity ore sourcing; it was pointed out that the heterogeneity of impurities in a single ore source, the differential effects on various trace elements of different smelting procedures, and the presumed inconsistency of impurity patterns within a single artifact, not to mention the reuse of scrap metal, rendered ore sourcing problematic at best (Craddock and Giumlia-Mair 1988; Northover 1989; Slater and Charles 1970; Tylecote 1970; Tylecote et al. 1977).

These criticisms appear to have been unduly pessimistic. While it is nearly impossible to isolate the ore source for the metal of a particular artifact (for a number of reasons ranging from fluctuations in the quantities of impurities in a single ore body to the exhaustion of ancient ore sources), one *can* often determine if a collection of artifacts came from the same ore source and/or were smelted the same way (Craddock and Giumlia-Mair 1988). In the present study, for instance, I use the results of trace element analysis as evidence that the heterogeneity of ore sources increased in the period 50 B.C.–A.D. 70; trace element analysis also shows that the copper used to make brass artifacts differed in source from the metal used to make copper and bronze artifacts. Thus both alloy composition and impurity pattern can become cultural markers (Hamilton 1991; Liversage and Liversage 1989; Waterbolk and Butler 1965), and provide clues to sociotechnical change.

In many areas scholars have discovered a precise correlation between composition, typology, and distribution, which provides a solid basis for studies of industrial organization and trade patterns. (Northover 1989 is an excellent example of this kind of study.) In addition, ore sourcing has also been done using lead isotope analysis rather than micro-impurities, with some success (Gale and Stos-Gale 1982; Stos-Gale 1989; Yener et al. 1991).

Concern that the inhomogeneity of ancient metals complicates analysis has also been exaggerated. Bulk segregation of elements occurs only rarely and mostly in elongated artifacts such as swords (Northover 1989). The researcher does need to take account of surface segregation of such elements as tin and arsenic, however (Tylecote 1985).

Studies of alloy composition can answer a number of questions, such as whether a certain alloy was preferred for a certain class of artifact, how much consistency the makers could maintain in metal production, and the effects of scrap metal circulation. Only by detailed study of the properties and characteristics of particular alloys, and the details of the particular metallurgical processes in use, can one obtain clues as to *why* a specific alloy was used.

Having established the compositional details, one can use them in the study of technological change and the transfer of technology, for example, by tracing the spread of new alloy types (brass, in this study), the appearance of which implies the acquisition and acceptance of new technological information. This is of special interest to anthropologists because the incorporation of new technology—in this study, the use of brass and the cementation method for manufacturing brass cheaply—frequently implies the creation of a new sociotechnical system, with concomitant adaptations in the structure of labor and the value of artifacts and processes.

Metallographic analysis

Another form of analysis in archaeometallurgy is metallography. Metallography is performed by cutting a small (at least 1–2 mm) sample from an ancient metal artifact, embedding the sample in an epoxy resin, grinding it flat, polishing it with increasingly finer diamond and alumina grit slurry until all scratches are gone, and finally etching it with one of many etchants. The etched sample is then examined under 50×–600× magnification (Scott 1991). What is seen in the microscope reveals the metallurgical history of the artifact, for metal retains the memory of everything that has been done to it; the microstructure is "a frozen slice of history" (Smith 1981:70). Metallographic examination allows us to tell the difference between cast, wrought, and annealed metal, the degree of cold working, heat treatment, and the inclusions in a metal.[1] Metallography is also useful in the study of joining, plating, and coating.

Metallography has been an adjunct in most archaeometallurgical studies, which have traditionally concentrated on compositional analysis and supplied only the occasional illustrative photomicrograph. Micrographs should be part of any analytical publication; this information goes hand in hand with compositional and archaeological data (Northover 1989). In Childs (1991a, b), one can see an excellent example of the proper incorporation of metallography into a metallurgical study; see also Buchwald and Leisner (1990), Rostoker and Dvorak (1990), and Scott (1991). Only by analyzing the working techniques can a full picture of an industry be put together.

Other techniques

Many other forms of analysis are also possible. Study of raw materials and archaeological remains of smelting and smithing structures in the ground has already been mentioned. In addition, one can study mining sites (Craddock 1989; Jackson 1980; Jovanović 1980; Rothenberg and Freijeiro 1976, 1980), and the residues of metalworking, such as slags (Bachmann 1982; Photos et al. 1985; Zwicker et al. 1985) and the residues that cling to furnace walls, crucibles, and molds (Blair 1992; Craddock et al. 1985; Glumac and Todd 1991; Photos et al. 1985; Zwicker et al. 1985). Analysis of these waste products can reveal the alloy, the means of manufacturing, and whether it was ingot metal or finished artifacts that were produced at a site.

A final valuable technique is hardness testing, useful in assessing degree of working, provided that the composition is known. Hardness in metals is defined as the metal's resistance to indentation, either of a small ball (Brinell test), or a diamond pyramid (Vickers test). For instance, the Brinell hardness of worked and annealed copper is about 50, while worked and unannealed copper has a Brinell hardness of 110 (Hodges 1976:207). (These numbers are simply points on a relative scale of hardness.) The Brinell numbers for (1) 50% worked and (2) 50% worked and annealed 8% tin bronze are 195 and 82, respectively (Scott 1991:83). Tensile strength, while commonly tested for in modern metals (Butts 1954), is seldom analyzed in ancient metals.

The characteristics of copper and copper alloys

Like all metals, copper has a crystalline structure. The crystals can take a number of forms, depending on the element and the temperature: face-centered cubic (FCC) and body-centered cubic (BCC) are two of the most common. Most metals are solid at room temperature, hard, and *ductile*, i.e., capable of being stretched or deformed under pressure. Metals can be bent or flattened, while ceramics and flints, for example, cannot.

FCC metals, among them copper, silver, gold, and lead, are particularly ductile.

In nature, copper appears in two forms: native and as an ore. Native copper is pure copper metal with only a trace of impurities; it is quite soft and, according to some metallurgists, "can be worked almost infinitely without cracking" (Smith 1967:28, as well as Weaver 1954). Most copper deposits are in the form of ores; the copper is bonded to other elements and must be *smelted*, i.e., heated in a reducing atmosphere, in order to separate the copper from the siliceous gangue.

Native and unalloyed smelted copper can be hardened in two ways. *Work hardening* involves subjecting the metal to pressure, as during hammering or drawing. The pressure causes deformations in the atomic planes of the metal; these deformations accumulate and prevent the planes from slipping over each other (this slippage is the cause of ductility). This can increase the hardness of the metal by two times or more, as seen in the results of the Brinell tests above (Scott 1991:82).

The second method of hardening involves *alloying*, i.e., the deliberate or inadvertent mixing of a metal with another element. Alloying copper with arsenic, tin, antimony, zinc, nickel or silver will harden it; alloying with lead will soften it. Alloying also reduces the melting point of copper, as well as improving its casting qualities (Weaver 1954).

Copper alloys

Pure copper is quite soft, ductile, easy to work, and difficult to cast, because of the gases emitted during solidification. The various alloys of copper differ somewhat in their properties. The first copper alloy in use was *arsenical copper*, which resulted from smelting high-arsenic ores. The presence of 2–4% arsenic will result in substantial hardening. Arsenical copper was not found among the analyzed Titelberg artifacts (Table 1).

Table 1. Alloys identified at the Titelberg: names and compositions

bronze	copper, tin
leaded bronze	copper, tin, lead
brass	copper, zinc
leaded brass	copper, zinc, lead
ternary alloy (gunmetal)	copper, tin, zinc
quaternary alloy	copper, tin, zinc, lead

In many parts of the Old World arsenical copper was later replaced by the tin-copper alloy *bronze*. Alloying with tin produces harder metal than alloying with arsenic (Scott 1991:82–83), and it is much easier to extract tin from its ore than arsenic from its ore. The Vickers hardness figure for work-hardened native copper is around 115, for work-hardened 2.6% arsenical copper, 150–160 (Scott 1991:82), and for work-hardened bronze (6–14% Sn), between 221 and 241 (Coghlan 1951).

The addition of tin up to 10–13% deoxidizes, toughens, and hardens the metal; more than 10–13% tin results in the formation of a phase in the metal that is quite brittle, so that, while the metal becomes harder, it is also increasingly brittle and prone to cracking if worked. Bronze can be cast more easily than pure copper, but the casting qualities are much improved by adding at least 3% lead; not only is the fluidity of the molten metal increased but the melting point is lowered (Weaver 1954). Craddock and Giumlia-Mair (1988) maintain, though, that any percentage of lead over 2% offers no advantage in casting; this is borne out, they state, by the consistently erratic levels of lead found in ancient artifacts. Lead is largely immiscible in copper; when added to copper it remains scattered as globules of free lead. The addition of lead results in softer metal, so a balance must be struck between improved casting quality and hardness.

The zinc-copper alloy *brass* is softer than bronze (Shrager 1969:225); the Vickers hardness of a cold-worked 30% zinc brass is 120–160, of cold-worked 12% tin bronze, 220 (Scott 1991:82). But brass, especially brass containing 30–38% zinc or less (alpha brass), is a good general-purpose alloy; it both casts and works well (Romanoff et al. 1954; Weaver 1954). Again, the addition of lead improves castability, though it decreases the work-hardening capacity of the alloy. Modern metallurgical practice restricts the amount of lead in copper alloys intended for wrought work to under 4% and preferably lower (Weaver 1954). In antiquity, lead was often added as a cheap filler, as well as to improve castability (Craddock and Giumlia-Mair 1988; Smith 1967).

The properties of the *ternary* and *quaternary alloys* vary with the percentages of the added tin, zinc, and lead. A common modern alloy is leaded red brass, with 85% copper, 5% tin, 5% zinc, and 5% lead; this is a general-purpose casting alloy. Modern practice with unleaded copper-tin-zinc alloys is diverse; a tin bronze with 2–20% tin and less zinc is a common foundry alloy, but a tin brass with over 6% tin and more zinc than tin is seldom used in modern foundries (Romanoff et al. 1954).

Fabrication and metallography

Copper-base metal can be manipulated two ways: by *casting* or by *working*, including *drawing*. To cast an object, one pours molten metal into a mold made of stone, clay, or metal, with vent-holes for gases to escape. The molds may consist of one piece, for simple shapes, or multiple pieces, for more complex artifacts. A false core of clay can also be used to create a hollow casting.

Fig. 1:
Dendritic structure in a copper alloy (8% tin). 50×.

Working (called *forging* when done to iron) involves either *cold working*, i.e., working carried out with the metal at room temperature, or *hot working*, carried out with the metal either red-hot or warmed to annealing temperature. Working is carried out by hammering, raising, turning, or drawing an object to the desired shape, using a variety of hammers, anvils, and lathes. Wire in antiquity was produced by drawing copper through progressively finer holes.

Whether cast or worked, most copper and copper alloy work in antiquity was carefully finished by *planishing*, or lightly hammering the surface to remove working traces, and then filing and polishing for a fine finish (Hodges 1976:76).

Both casting and working leave clear traces under the microscope. The basic microstructures that result from these methods are described below.

Casting of copper and its alloys. A cast structure results when metal is heated to its melting point and then cooled, usually in a mold. Ancient cast copper-base objects often have a *dendritic* microstructure (Fig. 1). In pure copper, dendrites are a result of crystal growth. In alloys, dendrites result from crystallization at the different solidification temperatures of different metals. Copper, for instance, solidifies at 1083°C and tin at 232°C. When molten bronze begins to cool, it often solidifies first near the outside walls of the mold. At any given moment during solidification, one finds a zone of solid copper-rich metal (because the copper will solidify first), a zone of tin-rich liquid metal, and between them an interface zone where liquid metal is in the process of solidifying. Because of the differential rates of solidification of the copper, the tin, and any other impurities, this zone is ragged and assumes a dendritic, or tree-like, shape. If the metal is cooled relatively quickly, as was true of most ancient metals, the whole mass of metal solidifies before the content can become homogeneous and thus the dendritic structure is preserved, with a compositional gradient from the inner regions of the dendritic arms to the outer surface of the arms. This visible compositional gradient is referred to as *coring* (Scott 1991:5). Outside the arms exists another phase of metal, which solidified after the arms. In a bronze, this will be a tin-rich phase. "A *phase* is any homogeneous state of a substance that has a specific composition" (Scott 1991:5). Under equilibrium conditions the nature of the second (or third, or fourth, phase) can be predicted from a *phase diagram*, but few ancient artifacts were produced under equilibrium conditions (ibid.)

According to Scott (1991:5), "dendritic structures dominate the world of ancient castings." Occasionally other kinds of segregation occur, such as normal segregation, when the alloying element is concentrated in the interior of the object, and inverse segregation, when the alloying element gravitates to the surface (ibid.). This latter phenomenon can be seen in some objects alloyed with arsenic, tin, antimony, and silver (Tylecote 1985).

Other information can be gleaned from the metallography of dendritic copper alloy structures. The smaller the dendrites, the faster the rate of cooling. Cast metals often show porosity or spherical holes, due either to dissolved gases emerging during cooling or to incompletely filled interdendritic holes (Scott 1991:5). The quantities of tin in a bronze can often be estimated from the appearance of a third phase, the $\alpha + \delta$ eutectoid.[2] This light blue, jagged-looking structure begins to form at tin percentages of 5–15%.[3] This eutectoid is brittle, and if there is too much of it the metal can be difficult to work. Careful annealing of bronze with a tin content of up to 14% can eliminate this phase, with great improvement of working quality. With tin quantities of over 14%, however, full homogenization is impossible (Scott 1991:15).

Working and annealing of copper and its alloys. The second method of fabrication involves working and annealing. Casting and working are by no means mutually exclusive; frequently (especially in this study) a cast

Fig. 2:
Fibula from the Titelberg (#207-72), worked and annealed 25% zinc brass. 200×. $NH_4OH + H_2O_2$.

object would be further worked and annealed. *Working* has already been defined (see pp. 14-15 above); *annealing* is the process of heating an artifact up to several hundred degrees C (usually around 500° to 800°C), but below the melting point of the alloy (Scott 1991:7). Working, as already described, produces distortions and strain in the metal; the crystals (or *grains*) become flattened. Annealing allows *recrystallization* of the grains; the strain deformations are relieved, the metal is softened, and the flattening of the grains disappears. The grains resume their normal energy-conserving hexagonal shape (Fig. 2). The softened metal is relieved of the threat of cracking and further working can proceed. A cycle of working and annealing can be repeated several times by the smith. Annealing also allows the cored dendritic structure to become homogeneous, i.e., without compositional gradients; this results in the disappearance of the coring and the appearance of equiaxed[4] grains. Frequently, however, annealing is not carried out long enough and "ghost dendrites" remain visible along with the newly recrystallized grains.

One unmistakable sign that both working and annealing have taken place is the appearance of *annealing twins*. Face-centered cubic metals such as copper recrystallize by twinning, which produces a mirror reflection plane within the grains. This results in the appearance of parallel straight lines, or twins, within the individual grains (Fig. 2).

Further working after annealing results in distorted grains, bent annealing twins, and *strain markings:* fine intragranular parallel or crosshatched lines.

The more heavily the artifact has been worked (deformed), the smaller the recrystallized grains will be. Recrystallization is faster, and therefore produces smaller grains, when the metal is very deformed, because the extra strain energy can be used to reorient the atoms; recrystallization is also faster at higher annealing temperatures (Hudson 1973:86).

To sum up, the basic structures that the metallographer of ancient metals sees, barring complications, are as follows (see also diagrams in Hodges [1976: 214–218]).

1. Dendrites going inward against the direction of heat loss. Cast, unmodified. See Fig. 1.
2. Dendrites flattened. Cast, worked.
3. Some grains, ghost dendrites. Cast, some annealing, no working. See Fig. 3a.
4. Equiaxed grains. Fully annealed. See Fig. 3b.
5. Flattened grains. Annealed, worked. See Fig. 3c.
6. Grains, annealing twins. Worked, annealed, possibly several times. Could also be the result of hot working, i.e., deforming the metal while it is hot. This was done rarely with copper-base alloys; it was frequent with iron. See Fig. 3d.
7. Grains, annealing twins, strain lines/bent twin lines/flattened grains. Worked, annealed, worked. See Fig. 3e.

The history of brass

Although zinc is one of the most common metallic elements in the earth's crust, in western Asia and the Mediterranean the use of zinc as an alloying element became routine only in the first century B.C. (There is evidence that it was available in India as early as the fourth century B.C. [Craddock et al. 1990].) The use of zinc to alloy copper offered several potential advantages; zinc is commoner than tin, lowers the melting point of the alloy as tin does, hardens the copper, and, in the proper percentages, gives the alloy a golden color. Only a few percent of zinc can color the alloy noticeably; modern gilding metal used for jewelry is composed of 95% copper and 5% zinc (Weaver 1954). The difficulty of extracting the metal from the ore, however, precluded large-scale use until the invention of the cementation process around 100 B.C. The source of the difficulty lies in the high volatility of zinc; it vaporizes at 917°°C, well below the temperature at which it could be smelted. Thus any attempt to extract the metal from the ore would result not in the metal puddling at the bottom of the furnace, as happens with copper, but with zinc leaving the furnace as a vapor (Craddock 1990).

Distillation and condensing of small droplets of zinc was possible, but does not seem to have become economic until the thirteenth–fourteenth centuries A.D. in India (Craddock et al. 1990) and Iran (Tylecote 1970).

The difficulty could be overcome in the cementation process of alloying, in which finely divided pieces of copper are mixed in a closed crucible with zinc ore (calamine) and charcoal. The zinc vaporizes at 917°C; pure copper melts at 1083°C. The interior of the crucible must be kept between these two temperatures—above 917° to allow the vaporization of the zinc, and below 1083° to prevent the melting of the copper, since the zinc moves into the copper by a process of diffusion and the surface area of copper in its liquid state is too low to permit significant diffusion. As the zinc content of the copper becomes richer its melting point drops, so the effective limit on the percentage of zinc in brass made by the cementation process is 28% (Craddock et al. 1980). Not all artifacts made of cementation brass will contain 28% zinc, of course; the proportion of zinc absorbed and the degree of penetration depend on the surface area of the copper, the purity of the copper, the amount of zinc ore, the temperature inside the crucible, and the length of time the crucible was heated (Caley 1964).

After maximum diffusion, the solid copper-zinc alloy is then melted and stirred in order to homogenize the metal, since the zinc will have only diffused into the outer portion of the fine copper pieces.

Early brass

Occasionally European or Middle Eastern artifacts allegedly dating to the Bronze Age have been found to contain up to 30% zinc (Caley 1964), though given their lack of provenience they are likely to be fake (Craddock 1978). Artifacts with several percent of zinc are more likely to be the result of accident than fraud. The most famous of these early zinc-containing copper artifacts are the 16 pieces from the Early Bronze Age tomb at Vounous-Belapais on Cyprus (Stewart 1950), which, according to Craddock's widely cited 1978 work on early brass, had zinc percentages of 1–8%. Later work by Craddock and Giumlia-Mair has revealed, however, that only one of these pieces had in fact ever been analyzed; when these authors did do the analyses, only the one originally studied contained any zinc. Craddock and

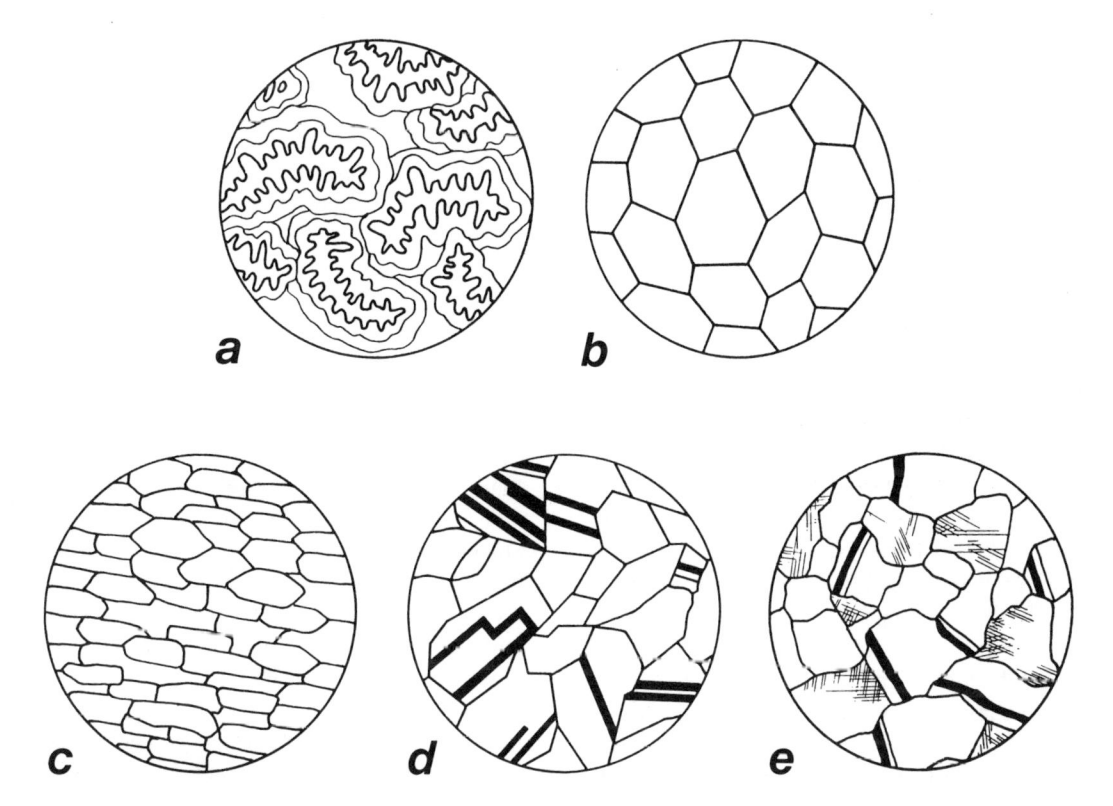

Fig. 3:
Microstructures in copper-base metal. From Scott (1991:7).
(a) Cast, partly annealed, showing ghost dendrites; (b) fully annealed, showing equiaxed grains; (c) annealed, worked, showing flattened grains; (d) worked, annealed, showing annealing twins; (e) worked, annealed, worked, showing annealing twins, strain lines/bent twins/flattened grains.

Giumlia-Mair (1988:320) politely refer to this as "a sad mixup." They assure us, though, that provenienced metalwork with several percent of zinc that was obtained from smelting zinc-rich copper ores does indeed exist; they cite a ring from Ugarit with 12% zinc and an axe from Beth-Shan with 6.5%; both date to ca. 1400 B.C.

The earliest deliberate manufacture of brass seems to have occurred in Anatolia, and is assumed to have been by distillation, both because of the relatively low levels of Zn and because slightly later descriptions of the use of "false silver" describe what seems to have been a distillation process (see below). This early use of zinc appears in a series of eighth–seventh century B.C. fibulae (brooches) from Gordion that were found to contain over 10% zinc and 6–15% tin; other objects such as vessels and siren attachments also had from 2 to 12% zinc (Young 1981). The zinc was probably added for the golden hue it lends the copper. Two well-authenticated Etruscan statuettes (one fifth century B.C. and the other third–second century B.C.) contain about 12% zinc (Craddock 1978). No well-provenienced pieces of pre-Roman Greek brass exist, though the only surviving piece of true metallic zinc in antiquity was found in excavations at the Athenian Agora in a context dating from the third–second centuries B.C. (Caley 1964; Craddock 1990). The rarity in Greece of other finds of zinc and brass must be due to the difficulties of distilling pure zinc.

Despite the rarity of pre-Roman brass in Greece, literary references to *oreichalkos* (Latin, *orichalcum*), or "copper of the mountain," appear in Greek literature from the seventh century B.C. on. *Oreichalkos* in later Greek writings means "brass," and this is probably its early meaning as well. From the description of Theopompus (fourth century B.C.), we know that in Asia Minor *oreichalkos* was made by combining copper with droplets of "false silver," which was almost certainly metallic zinc produced by an inefficient distillation process (Craddock 1978; Halleux 1973). It is notable that all these literary references speak of *oreichalkos* as being a most rare and precious metal (Halleux 1973), which, if the main source was the drop-by-drop distillation of vaporous metallic zinc from selected ores, would certainly have been true. The earliest reference (in the anonymous seventh century poem "Shield of Heracles") speaks of "shining *oreichalkos*, the glorious gift of Hephaistos"; Plato speaks of *oreichalkos* as being the second most precious metal of the Atlanteans. (See Craddock 1978 for a fuller list and discussion of the ancient references.)

To conclude, the use of brass before the first century B.C. seems to have been rare and sporadic, with the first regular use having been in Phrygia in Asia Minor. The Greeks considered it to be an exotic, highly valued metal, probably because of both its rarity and its resemblance to gold, and when a source is named it is in Asia Minor (Caley 1964; Craddock 1978).

Brass in the first century B.C.

When the cementation process for the manufacture of brass was invented is not certain, but it was probably near the beginning of the first century B.C. It is then that we start to see the first regular use of brass anywhere in or near the Roman world. Other clues also lead to the conclusion that the cementation process was being used then: the zinc percentages in these first century artifacts are usually in the range 19–28%, and never exceed 28%; in addition, these early brasses contain very little lead or tin, suggesting that the copper base was newly smelted. The presence of lead or tin in copper intended for cementation is deleterious since these elements lower the melting point of the copper and nearly eliminate the already slender temperature window wherein the diffusion of zinc can take place (Craddock et al. 1980).

Conveniently for the archaeometallurgist, the first use of cementation brass appears to have been for coins, which tend to be datable. The earliest brass coinage is, unsurprisingly, from Asia Minor—specifically, an issue of Mithridates VI of Pontus (90–75 B.C.), as well as coins from Pergamun (ca. 50 B.C.) and Phrygia (70–60 B.C.). As mentioned, judging by the high zinc content of the coins, as well as by the low tin content, it is fairly certain that they were made by the cementation method (Craddock et al. 1980).

The first coin issue of brass in Italy was of C. Clovius, produced for Julius Caesar in 45 B.C. (Crawford 1985); there are no dated artifacts of Roman brass earlier than this. It was with Augustus, however, that we see the first mass production of orichalcum coins, replacing the earlier highly debased copper/bronze issues. Around 23 B.C. Augustus issued two denominations in brass, *sestertii* and *dupondii*. *Asses*, *semisses*, and *quadrantes* were issued in pure copper (Carter 1971). Every succeeding Roman emperor issued brass coins, though the zinc content declined steadily from the first century A.D. to the fourth century (Caley 1964; Carter 1971; Carter and Buttrey 1977; Craddock 1978; Riederer 1974a). The orichalcum used for the coins of Augustus through Claudius (i.e., from 23 B.C. to ca. A.D. 54) seems, from its purity, to have been newly manufactured; the higher impurity percentages in later brass coins imply that scrap metal—old coins and other metal—were being melted and added to the stock (Caley 1964; Riederer 1974a).

Before the Titelberg material was analyzed, no brass artifacts other than coins had been recorded for the late Republican period of the Roman world, including adjacent Gaul (Caley 1964); Grant (1946:88) suggests that the alloy was under an Imperial monopoly, at least for the first few decades of the Empire. This of course did not prevent the manufacture of a few brass objects, as this study shows. The source of the metal was either remelted coins, as has been suggested for the late first century B.C. Celtic coins of Poitou, legend *Contoutos* (Allen 1980:35), and the issue of *Germanus Inditilli L.* of the same date (Crawford 1985:218), or areas in Gaul with a knowledge of brass manufacturing acquired either in Rome or in the East. Brass use expanded widely in the first and early second centuries A.D., perhaps because of its gold color. In addition, niello, a popular decorative material of the time, will adhere to brass but not to bronze (Craddock 1978). It has been assumed also that brass was used widely because it should have been much cheaper than bronze, once the cementation process was known (Grant 1946), but if this were true then it is strange that the zinc levels in copper-base metals declined steadily after the middle of the second century A.D.: it seems that little fresh brass was being manufactured and most artifacts, including coins and bronze pieces, were made of recycled scrap (Tylecote 1976:58–59).

The history of brass has been discussed in such depth because the data set under study here includes three pre-Conquest brasses, as well as several later first century B.C. fibulae. The three pre-Conquest brasses from the Titelberg date from before the time of Roman use of brass in coins. It is unlikely, therefore, that the early brass found at the Titelberg was of Roman origin. With the exception of the two Etruscan statuettes cited above, the Titelberg brass artifacts are the earliest to be found in Western Europe. It is probable that the source of the metal lay in Asia Minor, perhaps by way of the Gauls of Galatia. If the reputation of orichalcum as an exotic, precious substance survived into the early first century B.C., then the inhabitants of the Titelberg would have had more reason than simply the color for valuing this material highly. The brass artifacts of the Titelberg dated to after 45 B.C. could have come from Roman coins, artifacts of Asia Minor, or Gallic workshops.

Previous metallurgical analyses of Late Iron Age and Roman artifacts

A wide variety of Roman period copper-base artifacts has been examined, using a diversity of analytical methods.[5] These analytical methods include wet chemistry, ultraviolet spectroscopy, atomic absorption spectroscopy, X-ray fluorescence, and neutron activation analysis. Far fewer non-Roman Iron Age artifacts have been analyzed. The most important analytical studies, and the ones most useful for comparative purposes to this study, are summarized below.

Statuary bronzes

Caley (1970) collected already published analyses of Greek, Roman, and a few Hellenistic statues, all from museums and with very uncertain proveniences. Despite the purely stylistic dating, the patterns are clear: the pre-Hellenistic statues were of pure bronze (with an average tin percentage of 11%) and no added lead, while the Roman statuary bronzes usually had more lead than tin, and less tin than the Greek pieces. Unfortunately, however, there were very few Hellenistic samples, and the Roman statues could not be dated relative to each other. So the course of this changing technological recipe could not be traced. No zinc was found in any of these statues; Caley used the presence of zinc in Roman and Greek statuary as a sign of forgery (Smith 1970:55).

In contrast to Caley's findings, Woimant and Hurtel (1989) report the analysis of a single statue of a clearly Gaulish "god warrior," datable to approximately the beginning of the first century A.D. It was discovered in situ in a Gallo-Roman shrine in Oise, France. Using ultraviolet spectroscopy, Woimant took samples from 17 different places on the figure. Tin and lead were always below 1%; zinc levels ranged from 13% to 23%. This was therefore a pure brass.

The most comprehensive analytical study of Gallo-Roman figurines is in Beck and colleagues' (1985) study of a miscellany of copper-base artifacts from different areas of France. Of their total artifact set, 167 were figurines. Beck and her co-authors sought to answer questions such as: Can compositional results give provenience? Can workshops be localized? Are knowledge and techniques circulated? Is recycled metal traded widely? Do different workshops use different recipes?

Most if not all of the 465 artifacts come from museum collections, with considerable problems of provenience. Many are attributed to the region of the first known owner. No attempt was made to date the figurines. In addition, up to 40 of the 465 pieces might be fakes.

Nonetheless, patterns can be discerned. Using ultraviolet spectroscopy, it was determined that 92% of the figurines had over 2% lead, compared with 73% of the artifacts as a whole. The mean lead was 10%, but the amount was extremely variable and was associated with particular geographical regions. In contrast to Caley's results, 48% of the figurines had zinc, usually as part of a quaternary alloy. The mean zinc level was 10%, but, again, was variable. Both the varying proportions of the alloying elements and the results of lead isotope studies suggest that there were indeed regional "schools" or areas characterized by typical alloy recipes. Metal was indeed recycled, but largely within the same region. The alloys hint at a jealous preservation of workshop secrets, which Beck and her colleagues (1985) suggest is an inheritance from quarrelsome Late Iron Age times. Beck and her co-authors conclude that there was little movement of skilled workers or goods in Roman times, but a considerable diffusion of general technical knowledge.

Picon and colleagues (1966, 1967) and Condamin and Boucher (1973) have confusing results from their figurine studies. Their statues are from French museums, with poor provenience, and were analyzed by optical emission spectroscopy. In contrast to Caley, they found considerable lead in Greek and Etruscan figurines. They found no zinc in Greek and Etruscan statuary, but a class of grotesque figurines called "Alexandrine" (thought to have originated in Egypt) or "pseudo-Alexandrine" was characterized by 3–16% zinc. In Gallic statues, as well, there was zinc. The workshops in central Gaul and the Rhineland seem to have used more zinc than other Gallic areas. The researchers conclude that the use of zinc in figurines was not a general tradition in the Roman Empire, but was restricted to Gaul and Alexandria.

Vessels

Using atomic absorption spectroscopy (AAS), Stutzinger (1984) analyzed 61 Etruscan, Late La Tène, Campanian, and Gallo-Roman vessels from the collection of the Römisch-Germanischen Museum in Köln. The artifacts came principally from the area around Köln. The Etruscan, La Tène, and Campanian vessels were all of tin bronze, with an average of 10% tin. Some lead appeared in the Campanian and Gallic pieces. Pure copper, brass, and quaternary alloys appeared only in late Roman period vessels.

Beck and her colleagues (1985) also analyzed 78 cauldrons from France. The vessel bodies were invariably tin bronze, with the handles cast in leaded bronze.

Military fittings

Craddock and Lambert (1985), using AAS, analyzed 20 small copper-base fittings from the Roman military base of Xanten in northwest Germany, on the Rhine River. The small decorative pieces dated to the first century A.D. The researchers discovered that 18 of these fittings were of slightly leaded brass, with zinc levels of 16–24%. The other two pieces were of copper. They report that, based on a survey in Craddock (1978), 50% of Roman decorative metalwork and nearly 100% of the military decorative metalwork of the first century A.D. contained zinc. (The original source of these survey data is not named.) They suggest that the main sources of brass metal in the provinces were Roman coins and Roman military fittings.

Fibulae

The most important study of fibulae, or brooches, is the massive study of British fibulae undertaken by Justine Bayley and her colleagues (Bayley 1984, 1985, 1990; Bayley and Butcher 1981). They analyzed, using AAS and X-ray fluorescence, some 2,000 Late Iron Age and Roman period artifacts, two thirds of which were brooches. The fibulae came from known sites; they were also reasonably well dated by type. Analysis revealed that brass first appeared at the beginning of the first century A.D., some decades before the Claudian conquest of Britain (A.D. 43); these early fibulae appear to be imports from the Continent (Bayley 1990).

The brooches were made of a variety of alloys: brass, leaded bronze, copper, bronze, and quaternary alloys. There was positive correlation of alloy with fibulae type. The use of brass for fibulae was concentrated in the first century A.D. The virtual disappearance of brass in favor of leaded bronze, and, to a lesser extent, quaternary alloys, suggests to Bayley and her colleagues that brass, which may have been under Imperial control, was becoming less available and smiths turned to recycling and dilution of scrap with lead. She does not suggest, as did Craddock (1978), that this was caused by the superiority of quaternary alloy (leaded gunmetal) as a general-purpose alloy.

Coins

Because coins can usually be dated, many analytical studies have been done on them. For Celtic copper-base coins, there is the work of Gruel and colleagues (1979) and Rousset and Fedoroff (1985). For Gallic first century B.C. *potin*, or high-tin cast bronze coins, we have the work of Northover (1992), who collected recent analyses of a wide variety of coins, both British and Continental. His Gaulish and Helvetic coins had tin from 5% to 40%, with lead from 0.05% to 14%, though most were low in lead. British cast bronze had 13–37% tin and 0.05–15% lead. However, cast coins of the Durotriges, a British tribe, had considerably lower tin levels: 5–19%. These coin analyses will be mentioned again when discussing the possible function of the Titelberg workshop.

For Roman coins, useful studies are by Carter (1971, 1978), Carter and Buttrey (1977), and Riederer (1974a). These trace the changes in Roman issues from the reorganized copper and brass issues of Augustus (23 B.C.) to the debased currency of Gordion III in the third century A.D. The results demonstrate clearly that the zinc content of Roman brass coins declined steadily from the middle of the first century A.D. to the third century, at which point the coins contained virtually no zinc. Tin and lead took up the slack.

More sensitive analyses have been done than simple plotting of macroelements. Carter and Buttrey examined trace element patterns of copper-base coins of Augustus and Tiberius and discovered that the products differed in trace elements not only by mint but also by year. They also discovered that the copper used for brass coins contained more nickel than the copper used in the contemporary copper *as*,[6] indicating that the source of copper was distinctly different. Carter (1978) examined 180 Augustan *quadrantes* dating to 9–4 B.C., and determined that the trace element patterns correlated with the year of manufacture; an additional study done on Roman copper and brass coins demonstrated that the various Imperial mints from Spain to Antioch used metals of varying regional compositions (Carter 1971). He was even able to determine that a new nickel-rich copper ore source was exploited in 23 B.C. and A.D. 22–28.

Multiple types of artifact and debris

An illuminating study has been done at the site of the Gallo-Roman town of Alesia in France (Rabeison and Menu 1985). According to Pliny the Elder (*Natural History* 34), who lived from A.D. 23 to 79, Alesia was noted for its bronzework. Using ultraviolet emission spectroscopy, Rabeison and Menu examined artifacts and manufacturing debris from five workshop areas uncovered in excavation; most of the workshops dated to around the first century A.D. Two of the workshops had debris and remains of brass only, showing that brass was worked in specialized workshops. The ingots from Alesia were of pure copper, which indicates that raw material circulated without alloying elements. This would agree with Beck and colleagues' hypothesis about regional workshops with jealously guarded individual recipes for alloys.

There was considerable tailoring of artifact to alloy in the Alesian workshops: all four fibulae examined were of brass, all the fitting-harness pieces were of brass, as were pendants and appliqués. Rings and rouelles (small spoked-wheel shapes of uncertain purpose) were of leaded bronze, and vessels were of tin bronze. In the first century A.D., tin appears to have been rare and expensive; no artifact had more than 9% tin, and brass and leaded brass dominate the collection (Rabeison and Menu 1985). As we have seen from the other studies cited, the use of zinc declines in coinage after this century, arguing that zinc was not cheap and readily available. Though the average zinc level in other artifacts declines as well, the decline is not as dramatic as in the coinage, and the percentage of copper-base metalwork that contains zinc at all increases from 25% in the first century A.D. to 40% in the fourth century A.D. (Craddock 1978:14). Craddock (1978:11) maintains that zinc was consistently ten times cheaper than tin and half the price of copper, but he does not cite his source of information. The more frequent appearance of a zinc-containing alloy may simply reflect the wider use of scrap.

Other studies of miscellaneous Roman period objects have been done by Picon and his colleagues (1966), Riederer (1974b, 1983, 1984), Riederer and Briese (1972), and Riederer and Laurenze (1980). These are on museum material or on finds dredged from the Tiber River, and they confirm the tailoring of artifact to alloy in the Roman period. The choice of fabrication method—casting, cold working, turning—was always appropriate to the alloy chosen.

Craddock (1985), in his massive review of 3000 years of copper alloys, lists many analyses, but since he seldom even names the site, much less the precise chronology, his compilation is of little use to this project.

Summary of analytical work

Despite the chronological problems in these analyses, some conclusions about metal use and technological traditions in the Roman and Gallo-Roman period can be drawn. Roman period metalsmiths had very clear ideas about what copper alloy they or their customers thought appropriate for various purposes. Vessels and cauldrons were usually made of tin bronze. In late Roman times quaternary alloys were sometimes used for vessels. Military fittings—the small hooks, weapon mounts, helmet decorations, belt fastenings, and other paraphernalia associated with a soldier's kit—were mostly of unleaded brass, at least in the first century A.D. Again, we do not know why; one can only speculate that there was a deliberate decision to associate soldiers and coinage together as agents of the government.

Figurines are complex in their composition. Roman period figurines almost invariably contained over 2% lead, both to aid casting and as a cheap filler, but figurines found in Gaul contained considerable zinc as well, and the figurines found in central Gaul and the Rhineland had the most zinc of all. A single Gallic god-warrior statue was of pure brass, with no lead or tin. This seems to indicate that brass and zinc were most popular in certain regions of Gaul and Germania. This bears out Beck and colleagues' conclusions that workshop regions with special recipes for metal can be distinguished in the artifactual record.

Certain general trends hold true for all the coins in the studies: Celtic coins (both pre-Conquest and post-Conquest) were of leaded bronze; the first Roman brass coins were of high-zinc brass, with the zinc content progressively declining through the centuries. The main mints, however, all seem to have had particular traditions or recipes for their coinage metal; clearly there was no rigid standardization.

Various types of fibulae can be correlated with various types of alloys. Most fibulae of Gaul and Britain were made of brass in the first century A.D.; after that brass vanished in favor of leaded bronze and quaternary alloy. Again, the availability of zinc seems to have become restricted.

The trend of analyses is promising. Scholars are putting more effort into studying both chronologically and regionally specific artifact sets, as they try to ascertain what patterns of metal use existed at certain times and areas. Scholars are now trying harder to ascertain changes through time and space. The excavations at Alesia show patterning of workshops, and demonstrate that brassworking was carried out in different workshops and city quarters than bronze- and copperworking. Bayley's work on fibulae demonstrates clear changes through time in alloy use in one artifact category.

These studies on patterning in time and space would sound rather routine in most archaeological contexts, but given the provenience problems of most Gallo-Roman and Roman artifacts, and the long-standing assumption that the label "Roman" was enough to ensure uniformity, they show that Gallic and Gallo-Roman archaeometallurgy has come a long way, and promises much more.

Notes

1. *Cast* metal is created by pouring molten metal into a mold and allowing it to cool; *wrought* metal has been hammered or deformed in order to fashion it into a shape; *annealed* metal has been reheated to below the point of melting. This annealing often results in the recrystallization of the microstructure. To *cold-work* means to hammer or deform metal at room temperature.

2. A eutectoid is a structure that results from the decomposition of a single solid phase into two finely dispersed solid phases. In tin bronzes, the eutectoid is made up of alpha, a copper-rich solid solution of tin in copper, and delta, an intermetallic compound of fixed composition, $Cu_{31}Sn_8$ (Scott 1991).

3. Though in equilibrium conditions this eutectoid should not appear in alloys with less than 14% tin, in reality it can appear in alloys with as little as 5% tin, provided cooling is rapid. The faster the rate of cooling, the lower the tin percentage required for the appearance of the eutectoid (Michael Notis, pers. comm.).

4. Equiaxed grains are of equal dimensions in all directions. These hexagonal shapes are a result of the maximal conservation of energy (Scott 1991:140).

5. Chase (1974) conducted a test of the comparability of compositional findings of various laboratories and methods. He reduced two ancient copper-base objects to powder and sent samples to 21 different laboratories for compositional analysis; the results demonstrate that agreement between laboratories and methods was "fairly good," but with discrepancies. This is a point to keep in mind when assessing the diversity of studies described in this section.

6. *Asses* and *quadrantes* were small copper coins, used primarily for small change and votive offerings (Carter 1978).

4

THE ARCHAEOLOGICAL AND HISTORICAL BACKGROUND: GAUL AND THE ROMAN EMPIRE

> The Celtic feasts described in Posidonius or the Irish tales, show us swaggering, belching, touchy chieftains and their equally impossible warrior crew, hands twitching to the sword-hilt at the imagined hint of an insult . . . wiping the greasy moustaches that were a mark of nobility, "moved by chance remarks to wordy disputes . . . boasters and threateners and given to bombastic self-dramatization." (Piggott 1965:229)

This chapter will present a background summary of the archaeology and history of northern Gaul, the area called by the Romans *Gallia Belgica*, during the two centuries before and after Christ. First, Gaul as a whole will be discussed (see Fig. 4), then northern Gaul.

A brief chronology is presented in Table 2, though some items in it need amplification. The La Tène chronology presented here is highly simplified; one of the difficulties of archaeology in Europe is that scholars in different countries have invented different terminologies for the same time divisions and are loath to abandon them. The designations La Tène I, II, and III are French; one can also use Early, Middle and Late La Tène. In the German Rhineland one uses La Tène A, B, C, and D, with La Tène D corresponding to La Tène III and La Tène C corresponding to La Tène II. In the Netherlands and North Belgium one simply uses Middle Iron Age to designate the Early La Tène and Late Iron Age to designate the later periods (Roymans 1990:7). The La Tène designations are not used to refer to the period after around 20 B.C., though characteristic La Tène material was still being made.

There are two main sources of evidence for Late Iron Age transalpine Europe: archaeology, and Greek and Roman writers such as Posidonius (135–50 B.C.) and Julius Caesar (100–44 B.C.). Though some transalpine Europeans were literate in the pre-Roman period, only a few inscriptions have survived (Kruta 1991).

The vexed question of ethnicity

Archaeologists have spent an inordinate amount of time on the question of the ethnicity of the Gallic tribes, particularly in northern Gaul. Caesar mentions three groups in Belgic Gaul: the Belgae, the Germans, and the Celts. It has been commonly assumed that these groups were "ethnic units," but it is not clear just what the

Table 2. Chronological chart: Gaul and the expansion of Rome

Iron Age Gaul	La Tène I	450–300 B.C.
	II	300–100 B.C.
	III	100– 20 B.C.
	Early Imperial/ Gallo-Roman	20 B.C.–A.D. 70
Roman Republic/ Empire	The Triumvirates	60–27 B.C.
	Augustus	27 B.C.–A.D. 14
	Tiberius	14–37 A.D.
	Caligula	37–41 A.D.
	Claudius	41–54 A.D.
	Nero	54–68 A.D.
	Gallic Revolt	68–69 A.D.
	Three emperors Galba Otho Vitellius	69 A.D.
	Vespasian	69–79 A.D.

Roman expansion in the West

By 120 B.C.:	into North Italy (Cisalpine Gaul), Spain (Hispania), Southern France (Gallia Narbonensis)
58–51 B.C.:	into Gaul (Julius Caesar)
A.D. 43–47:	into Britain (Claudius)

names *Belgae* and *Germani* mean. The Romans used the terms loosely, so that we do not know if in fact they referred to ethnic groups at all (Roymans 1990:13–14).

Much of the difficulty lies in the confusion of ethnicity, language, and tribe. The first certain mention of the term "Celt" or *Keltoi* was by Herodotus in the mid-fifth century B.C. The term appears to have been of Celtic origin, and suggests, since he uses it to refer to an entire "nation" (Herodotus 4.287), a sense of common culture or origin. (*Galatae*, or Gauls, is a variant term.) How far back "Celticity" extends is unknown; Celtic place names and the Iron Age archaeological cultures called Hallstatt C and D, and La Tène (750 B.C. to Roman Iron Age) do coincide. Hallstatt and its later development, La Tène, arose in Central Europe; Wightman (1985) suggests that it was subsequently spread by warlike Celts who became military aristocracies. In support of this, she mentions that in the Hunsrück-Eifel area around the Moselle valley (near the Titelberg), some pre-Celtic place names have been preserved. In other areas, such as Champagne, the degree of "Celtization" was greater (Wightman 1985:10). Other scholars, pointing out that there are no material culture breaks from the Late Bronze Age through the Iron Age, would trace the Celts back to the late second millennium B.C. (Coles and Harding 1979:336–337; Drinkwater 1983:9). The issue is essentially unresolvable.

The Early Iron Age coincided with (or the acquisition of iron caused) a great expansion of Celtic peoples, seen in the appearance of Hallstatt artifacts in Gaul, Spain, and Britain around 750 B.C.; the Celts also appear as fearsome and barbaric raiders in the Roman and Greek histories from the fourth century B.C. on. In Caesar's *De Bello Gallico*, he describes a tripartite Gaul, where the Gauls proper lived between the Garonne and the Seine/Marne Rivers and the Belgae lived north of the Marne. Caesar considered the Belgae different in both language and culture from the Gauls (Caesar 1.17), though it is now concluded that Caesar incorrectly

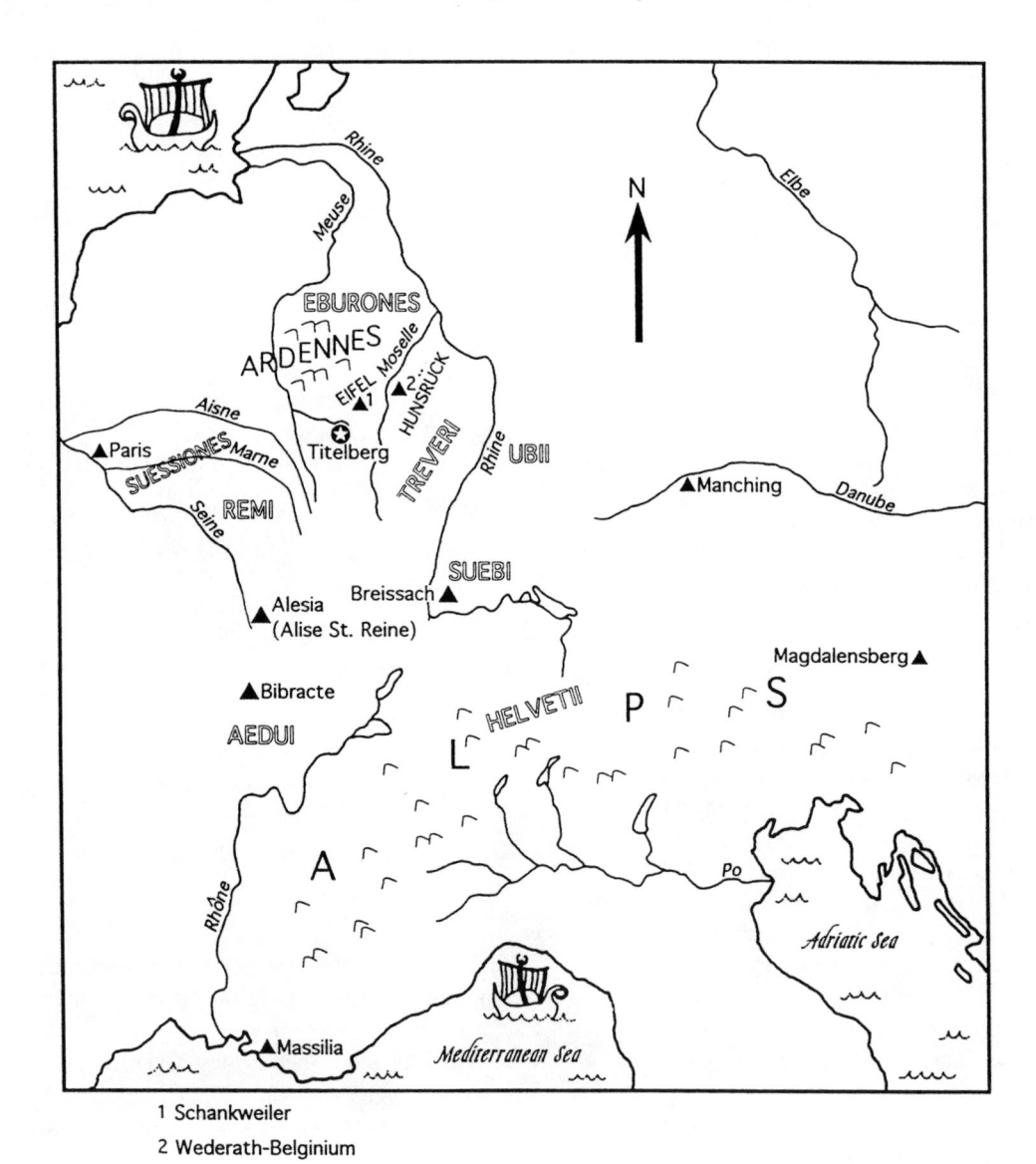

Fig. 4:
Pre-Roman Iron Age central and western Europe, with selected tribes.

lumped together all the north Gallic tribes as *Belgae*. The true Belgae were confined to Picardy and Upper Normandy, and almost certainly spoke a Celtic language (Roymans 1990:12).

According to Caesar, the Rhine was a cultural border; Germans lived to the east, Belgae and Gauls to the west. This contradicts his other assertion that the Belgae were of Germanic origin. In addition, he also refers to a number of tribes such as the Eburones as *Germani cisrhenani*, or "Germans on this side of the Rhine." The modern use of the term "German" is linguistic, but it seems unlikely that Caesar was using a linguistic definition for "Celtic" and "Germanic." Roymans (1990: 12–13) suggests that Caesar took the name of a relatively small group of tribes on both sides of the northern Rhine, the *Germani*, and extended it to all the transrhenine groups. Tacitus (ca. A.D. 55–120), in his *Germania*, notes that this broader use of the term was a recent one. At the time of Caesar's conquest, the people who actually spoke what we now call Germanic languages inhabited southern Scandinavia and central and north Germany; only later did they move into the area already called *Germania* and assume its name (Roymans 1990:13).

Thus it is impossible to state exactly what *Belgae* and *Germani* mean. Roymans (1990:14) suggests that "they refer to tribes which maintained certain sociopolitical and/or ceremonial relations with each other, but concrete evidence for this is lacking." It is equally impossible to assign the Treveri to one of these groups. According to Caesar, they claimed a Germanic ancestry, but we know from other evidence that they spoke a Celtic language.

Because these ethnic issues cannot be resolved, and in fact make little difference to this study, in this paper the terms "Gauls" or "Celts" are used interchangeably simply to refer to the inhabitants of Gaul and Belgic Gaul.

Pre-invasion social structure

The Celts of Europe were divided into tribes, which in some cases could contain over one hundred thousand people (Wells 1984:171). (In this context, "tribe" does not mean that social entity described by Service [1971:99–132]; the term is simply traditional. To avoid the confusion, some authors also use the Latin term used by Caesar, *civitas*.) Most of these tribes or *civitates* are classified by anthropologists as chiefdoms, though many scholars have maintained that in the late second and early first centuries B.C. a number of adjacent *civitates* in central France and Switzerland were developing into archaic or primitive states (Bintliff 1984; Crumley 1974:vii–viii; Nash 1976, 1978, 1985; though see Ralston 1988 for a counter-argument), primarily on the basis of the development of political-economic classes (Crumley 1974:vii). The classical authors agree on the stratified nature of Celtic society; Caesar describes a tripartite society with a class of warlike, oppressive, and wealthy nobles; a class of priests, scholars, and adjudicators known as Druids; and a much larger class of near-servile commoners. The writers do not mention a merchant class, but Crumley (1974:70–71) has hypothesized, from archaeological evidence at French *oppida* (Caesar's word, loosely meaning "towns"), the presence of a middle class of merchants and traders in central France. The commoners were apparently connected to the nobles by a patron-client relationship (Eisenstadt and Roniger 1980), the nobles providing protection and material goods and the commoners political and military support and probably agricultural products as well (Bintliff 1984; Crumley 1987).

Urbanization and production

The putative rise of the state in central France coincided with a remarkable increase in the quantity, variety, and standardization of material objects, particularly those made of iron. From the archaeological evidence, it seems fairly clear that there was more than one level of industrial production:

1. a hamlet or village level of ceramic and iron and bronze production (either smithing or smelting) by farmer-craftworkers (Wells 1984:146);
2. at least some itinerant craftworkers, as in Gussage All Saints in southern Britain, with its single season of bronzeworking (Wainwright and Spratling 1973);
3. small sites devoted mostly to the production of iron (Wells 1984:146); and
4. the *oppida*, at the top, which seem to have functioned, among other things, as production centers for coins, glass, bronze, textiles, and notably, iron, the quality of which had a high reputation in, and was exported to, the Mediterranean (Collis 1984a:156).

There is archaeological evidence that some of the *oppida* may have specialized in particular products; for example, iron production at the Magdalensburg in Austria, amber working at Staré Hradisko in Moravia, and glass working at Breissach on the Rhine River (Collis 1984b:146).

There is some dispute about the nature of the *oppida*, of which the Titelberg was an example. The classical authors Caesar and Suetonius use the word rather broadly to refer to towns and hillforts. The modern archaeological definition, at least as used by Collis (1984b), includes massive fortifications, large size (25–1000 ha), and an occasional or permanent population of several thousand (Collis 1984b:6; Wells 1984:165–166). Many of these *oppida* were located on or near easily worked sources of iron ore, others were near salt mines, graphite-rich clay, or on major trade routes, and many show evidence of considerable craft production. Some had areas of dense settlement, but all seem to have included large open areas, possibly used for agricultural pursuits and as refuge areas in times of war.

In Bibracte and Manching, two *oppida* that have been recently and extensively excavated, industrial quarters have been discerned (Collis 1984b:131; Crumley 1974:72). Collis (1984b:132) suggests a twofold organization, with craftworkers attached to aristocratic households in one quarter and small businesses owned by individuals along the major roads. Independent workers would still likely be clients of nobles, the distinction between "attached" and "independent" craftworkers resting on the kinds of things they made (Earle 1981).

The independent specialists, both within and without the *oppida*, would be producing utilitarian or small decorative goods—tools, cooking vessels, basic furniture, fibulae, and hairpins—for unrestricted local or regional use. The attached specialists would be under the direct control of the elite, "because control over production translates into straightforward control over distribution" (Costin 1991:11); the noble would decide where and to whom (clients, fellow nobles, the general public or independent retailers) the goods would go. This control would be strengthened and symbolized by the location of craftworkers in the aristocratic properties. They would produce fine weapons and luxury goods, "the emblems of power and prestige" that only a small portion of the population could obtain (Brumfiel 1987; Costin 1991).

Nonetheless, we know little about the degree of attachment of craftworkers to nobles, the size and flexibility of the work units, and the intensity of production in the Iron Age, and not a great deal more in the Gallo-Roman period. We do know from excavations at the cemeteries of Wederath-Belginium (Haffner 1989) and Schankweiler (Ludwig 1988), both in the region of the Titelberg, that rural and urban commoners had access to Roman luxury imports. Though the classical sources, as well as much later Irish writings (assuming that they are relevant for this period), do mention patron-client relations among the Celts, and it is reasonable to assume that many craftworkers were under some kind of noble domination, we have no information about how patronage worked in a craft setting. In this situation, the categories of "attached" and "independent" craftworkers found in the literature (Brumfiel and Earle 1987; Earle 1981) seem artificially dichotomous. If, for instance (in a situation extrapolated from noble/farming client relations found in the early medieval Irish law codes), the patron were supplying the smith with raw material and tools in return for a fixed percentage of the product or profit, with the rest of the goods produced being sold or bartered independently, the client would seem to be both independent and attached (Bernard Wailes, pers. comm.).

Although there is no evidence as to what the open areas inside the *oppida* walls were actually used for, Bintliff (1984) and Wells (1984) think that they were used for crops or pasture, and that even the allegedly full-time craftworkers of an industrial center would have spent part of their time raising their own food. This hypothesis has implications for the scale of organization and degree of attachment. Following Brumfiel (1987), one can argue that such a generalized strategy arises particularly in producers of utilitarian goods when a supply of market or imported food cannot, for a number of reasons, be relied upon.[7]

Despite the considerable evidence that many *oppida* were centers of industrial production, there is little archaeological or textual evidence that the *oppida* were administrative or ritual centers. No central places or structures have been identified in any *oppidum*, and only a few have been demonstrated to have *Viereckschanzen*, the square ditched enclosures inside of which are often shafts containing votive deposits. Many Roman-period temples were built on top of Iron Age religious areas (Collis 1984b:106), but many Gallic shrines were located in groves out in the countryside, so there is no reason to assume that the shrines in the *oppida* were more important than any other.

The role of metal luxury goods

Gallic society was both hierarchical and competitive, and one of the most usual means of expressing status was through the display of precious metals. Gold was the most highly valued, but silver and bronze, and later, brass, were also honored. The first Gallic coins were of gold and imitated the Macedonian *stater* of Phillipus II; these appeared around 250 B.C. (Allen 1980:4). Gold coins disappeared after the Conquest, as did most of the gold in Gaul; Caesar pursued a firm policy of assessing fines, despoiling sanctuaries, and general looting.

Metals were used for ornaments, cherished by both men and women (Roymans 1990:128), and for horse-trappings and vessels. Metals were used, too, in payments to clients, diplomatic gifts, votive offerings, dowries, and ransom (Roymans 1990:129–130). We can conclude that metal, especially gold, silver, and copper-base metal, carried a heavy symbolic weight to the Iron Age Gauls.

Northern Gaul

The tribes of northern Gaul had, on the whole, a less complex sociopolitical structure than the tribes in central

Gaul. In general, Gallic tribes, or *civitates*, were segmentary structures integrated by kinship, clientage, and territoriality. In the most northern part of Gaul, in contrast, "the *civitates* were often little more than a loose federation of a number of *pagi*, and there was little status differentiation within the broad group of regional chiefs" (Roymans 1990:261).[8] In the southern part of northern Gaul, especially in the territories of the Treveri and the nearby tribes the Remi and the Suessiones, the increasing competition between the chiefs and nobles "resulted in the emergence of a high-level elite, which possessed great wealth and had a large following of clients" (Roymans 1990:261). The power struggles between the two greatest Treveran nobles, Indutiomarus and his son-in-law Cingetorix, are documented in Julius Caesar's *De Bello Gallico* (5.1).

The great majority of the population was engaged in farming, especially the cultivation of cereals and the raising of cattle, pigs, and sheep. In the southern area of northern Gaul, there is a diversity of settlement. Little work has been done on excavating small rural settlements in the Treveran area, but a few unenclosed settlements elsewhere in northern France have been investigated; these are composed of large rectangular longhouses with pits and numerous granary-like structures (Roymans 1990:190). More important are the enclosed settlements, surrounded by ditches, that were found by Agache and Haselgrove in their intensive survey of the Aisne Valley (Agache 1978). They appear to have been the "fundamental component of the settlement system" (Roymans 1990:191); some contained only one large longhouse with ancillary buildings; others contained a complex of living structures. From the cemetery remains, a hierarchy of wealth can be inferred to exist among these settlements, larger ones of which appear to have been denoted by Caesar's term *vicus* (Roymans 1990: 190–192).

The area of Belgic Gaul contained fewer and smaller *oppida* than southern and central Gaul; the Titelberg at 43 hectares was one of the largest. (For comparison, Bibracte in central Gaul is 135 ha and Manching in Bavaria, 350 ha.) Nonetheless, the tribes of the Treveri, the Remi, and the Suessiones each possessed a number of fortified *oppida*. Caesar notes that the Suessiones had twelve *oppida*; they all fled into one upon his invasion, and defended themselves there. This only occurred in case of emergencies; in ordinary conflicts it is likely that each fortified place was defended by troops from its own *pagus*. Northern Gallic *oppida*, like southern *oppida*, have yielded evidence for religious worship, food storage, craft production, and areas of dense occupation (Roymans 1990:200–201).

Pre-invasion contacts between northern Gaul and the Roman Empire

Trade between the Mediterranean and Belgic Gaul flourished during La Tène I. Numerous Etruscan bronze vessels are found in graves in the area between the Moselle and the Rhine, as well as in the Marne valley in France; presumably they were traded for raw materials and slaves. Dietler (1989, 1990) suggests convincingly that the Celts were interested only in acquiring wine and its attendent equipment; this was by no means a full-scale trade. The trade ceased during La Tène II, roughly the period of Etruscan decline, and resumed in La Tène III, the period of Roman expansion and imperialism. The commonest and most chronologically significant Mediterranean artifact type found in La Tène III contexts is sherds of Dressel I amphorae, used to carry wine (Fitzpatrick 1985; Roymans 1990; Tchernia 1983). Pre-Conquest written sources mention both the notable Gallic fondness for wine and the profits to be realized from trading it (Tchernia 1983). Though most of the amphorae remains are concentrated in the south and central part of Gaul, concentrations are also found in the territory of the Treveri and the Aisne/Marne valley (Demoule and Ilett 1985). Dressel Ia sherds date from the last part of the second century B.C. to ca. 70–50 B.C. Presumably the trade was concentrated in the area of the Treveri because the main trading routes, the Rhine and the Moselle, ran through or by their territory, and they had the surplus wealth or slaves to trade in return for the wine. Italian bronze vessels were also imported, and are found in the same pattern of distribution as the amphorae sherds (Roymans 1990:151).

By the time of the conquest, Roman traders were common in Gaul. Caesar even used their difficulties with tolls as a pretext for opening battle in 57–56 B.C., and he also used them as spies (Roymans 1990:161).

Though excavations have documented the widespread presence of Roman luxury goods such as wine amphorae, fine ceramics, and bronze wine services in the period before and several decades after the Conquest, this, especially in central and northern Gaul, was no more than a thin layer of prestige imports on a consistent foundation of native material culture. Roman material culture was slow to appear in most contexts; at the Titelberg, Celtic style coins (in silver or copper-base) continued to be produced until ca. 25 B.C. (Weiller 1977), and until 30 B.C. Roman artifacts were few. The main road there was paved in Roman style and houses rebuilt in stone around A.D. 1; the production of native Gallic ceramics dwindled, but never fully ceased (Rowlett, pers. comm.). In the general area of the Treveri, cemeteries display an increasing number of Roman and imitation Roman artifacts, but most of the grave goods of the commoners stay La Tène in style at least until ca. A.D. 50, and the grave style itself remains unchanged (Hamilton 1995). The two rich graves at Goeblingen-Nospelt, 17 km from the Titelberg, dating from 50 B.C. to 30 B.C., contained only one Roman import, a Dressel 1B amphora (Metzler 1991; Thill 1967). Trier, later to become the principal city of the Treveran region, was founded in Augustan times, but did not look like a developed Roman city until the second century A.D. (Wightman 1971:75).

It is clear that the penetration of Roman material culture was relatively slow and sporadic in the north of Gaul. There was evidence of interaction with the Mediterranean before the Conquest, and the archaeological picture of interaction did not change significantly until several decades later. Indeed, the Titelberg reached the height of its native prosperity around 30 B.C. The thin overlay of Roman items was acquired for its prestige value, but there is no reason to postulate the presence of Roman artisans or techniques in northeast Gaul prior to Augustan times (27 B.C.–A.D. 14) at the earliest.

The Roman conquest of Gaul

The political history of the early Roman Empire may be familiar to readers; suffice it to say that Republican Rome in the first century B.C. suffered a breakdown in traditional modes of political power and was a battleground for ambitious men, best described as "warlords" (Drinkwater 1983:14). In 60 B.C. the three greatest contestants, Pompey, Crassus, and Julius Caesar, joined together in an unconstitutional concordat called by historians the "First Triumvirate" (see Table 2 above for chronology). Like most such alliances, this one was unstable, and Caesar, lacking the financial resources of Crassus and the military resources of Pompey, sought a governorship whereby, through conquest, he could win glory and riches. The governorship of *Gallia Narbonensis* (Provence) and Cisalpine Gaul was open, and he took it. He then seized the chance offered by a planned migration of the Gallic tribe of the Helvetii out of Switzerland to take his armies north and annex, over the next seven years, an area comprising some 535,000 sq km. Caesar left his account of the wars, along with much ethnographic information concerning the Gauls, in his *De Bello Gallico* (The Gallic Wars). It was, of course, a propaganda document, written to justify his actions and magnify his reputation back home, but, unlike the work of most other classical writers on the Gauls, it was based on firsthand observation, along with unattributed borrowing from Posidonius, an earlier ethnographer (Tierney 1960), and if used cautiously is a very valuable source of information (Crumley 1974; Nash 1976). Despite Caesar's self-justifications, it is clear that he picked quarrels with the Gallic tribes on the flimsiest of pretexts, with conquest and booty his major aim, and was very considerably assisted by the chronic disunity of the Gauls (Drinkwater 1983:14–17).

Political events after the Conquest

After Caesar crossed the Rubicon in Italy, he never returned to Gaul, and Gaul, politically at least, was left largely to its own devices. Three colonies of discharged legionaries were founded on the borders of the conquered territory in the 40s B.C., but by and large Caesar preferred to work with the already existing tribal leadership. Rome sought to use rather than destroy the surviving Gallic political structures, buying the loyalty of local leaders with status-enhancing gifts and concessions and leaning upon friendship treaties with two or three influential tribes. Taxes were relatively light, and trade and the production of silver and copper-base coinage continued. (The gold disappeared into the folds of Caesar's toga.) Nonetheless, the ever-present threat of the Germans and the possibility of Gallic revolt led to continued military occupation and martial law. Ironically, the army, that instrument of Gallic subjugation, was also a potent force for Romanization. Young male Gallic aristocrats, seeking glory and fortune in their traditional martial way, joined the Roman army as auxiliaries and emerged years later as Roman citizens, traveled, at least partially Romanized, and influential in their own communities (Drinkwater 1983:19).

Caesar's assassination in 44 B.C. led to a free-for-all power struggle that his grandnephew Octavian finally won in 27 B.C. Once he became Augustus, he took a personal interest in the affairs of Gaul, building military highways, ordering a census, and setting up a permanent structure of government and administrative seats, of which Trier (Augusta Treverorum, the capital of the Treveri) was one (Drinkwater 1983:20–25). The military roads, following Roman custom, were laid out with a total lack of regard for native centers of population (Dowdle 1987). There is no evidence of deliberate Roman depopulation of an *oppidum* (Collis 1984b:175), but very few remained inhabited long into the Roman period. He also attempted the conquest of the Germans across the Rhine, with a singular lack of success. The campaign was abandoned in A.D. 16, but the Germans were to prove a continual problem and legions were permanently garrisoned along the western Rhine bank. The economy and administration of Belgica were significantly affected by this perpetually consuming military presence (Drinkwater 1983:19–21).

Though Augustus, and after him, Tiberius, continued Caesar's policy of working through native elites as far as possible, there was still unrest, exacerbated by the corruption of Roman officials and the financial demands of the German war. In A.D. 21, there was a revolt led by Julius Florus of the Treveri and Julius Sacrovir of the

Aedui, both highly Romanized aristocrats; the revolt ended in failure, with little wholesale tribal support, but was symptomatic of Gallic uneasiness. In A.D. 48, Claudius allowed Gallic leaders to become members of the Roman Senate, which was not only a recognition of the increasing Romanization of the Gallic provinces but also of their prosperity (Drinkwater 1983:37).

His successor Nero's extravagances and neglect of Gaul led to the great Gallic Revolt of A.D. 68–69, which in fact was not a nationalistic revolt but a rebellion led in the name of Rome by Julius Vindex, a Romanized Gallic magistrate. This suggests that the Gauls considered themselves so firmly part of the Empire that they were concerned not with freeing themselves from it, but with reforming it. Julius Vindex in fact issued an invitation to Servius Sulpicius Galba, governor of Tarraconensis in Spain, to assume the imperial throne.

Vindex's revolt, however, was perceived by the Roman legions on the border as a native revolt, and Vindex's Gallic troops were wiped out. But the wildness of Nero's reactions provoked his many enemies in Rome to act against him, which resulted in his death. The Empire was offered to the elderly Galba, who accepted. This set the stage for a year of political upheaval in which three men in turn wore the purple and three men in turn died for it; the winner was the fourth contender, Vespasian, the hero of the Jewish Wars (Drinkwater 1983:44).

The demoralization of the Roman legions along the Rhine and the sufferings of Gaul during this period of civil war resulted in a successor to Vindex's revolt, a true nationalistic uprising led by the Batavian Julius Civilis, the Treverans Julius Classicus and Julius Tutor, and the Lingonian Julius Sabinus. They had early success; the Rhine troops were forced to surrender, and they acquired substantial German support. But their attempts to set up an *Imperium Galliarum* ultimately failed when the combined Gallic *civitates* rejected the revolt; their support indeed was mostly confined to northern Gaul. Their forces were broken by the Roman general Cerealis at the battle of Trier in A.D. 70 (Drinkwater 1983:47).

Vespasian did not exact widespread retribution in Gaul for the revolt, but it is likely that some punishment was handed out, particularly in the areas around the Treveri. Estates were confiscated; inscriptions show changes in patronage patterns. This retaliation may explain the destruction of the mint/workshop at the Titelberg ca. A.D. 70.

Belgica after the Conquest

The political boundaries of Belgic Gaul were based firmly on the communities of the pre-Roman period (Fig. 5). A few *pagi* were switched to other *civitates* and the Treveri lost their *pagi* on the east side of the Rhine. Despite this, the Treveri, as a large and strategically situated people, were initially exempted from tribute and prospered, as the excavations on the Titelberg show (Wightman 1985:56).

The administrative structure was based on treating the sprawling *civitates* as if they were Mediterranean city-states. This was not easy; Italian city-states were only a few hundred square kilometers; Gallic *civitates* could cover 10,000 sq km. Nonetheless, capital cities were appointed or founded and semi-Roman institutions set in place.

In the Roman colonies in Gaul, settled by Roman legionaries, standard Roman forms of city government prevailed. The populace elected a body of magistrates, who were supervised by *duoviri*. These *duoviri* had supreme judicial and administrative authority after the provincial governor, judging cases involving non-Roman citizens and sending citizen cases to the governor. *Quaestors* were financial officers; *aediles* maintained the public peace. The situation in the Gallic *civitates* varied considerably. Though they were conquered provinces, and had no legal right to govern themselves in this Roman pattern unless they were granted full colonial status, many of the Gallic nobility imitated the Romans in their administrative structure. The results tended to be unpredictable, with wide variations and adjustments to local circumstances; by no means did all *civitates* have *duoviri* and *aediles*. Rome did not appear to care about the specific administrative structure as long as the taxes were collected by the magistrates and the peace was maintained (Drinkwater 1983:107–109). Whatever the official structure, these *civitas* offices were held by native aristocrats, who received Roman citizenship after they left office (Wightman 1985:56–57).

They were loosely supervised by a procurator, a Roman official appointed to each province to supervise financial affairs (Wightman 1985:63). Each province was headed by a Roman governor. There was also the *concilium Galliarum*, the council of all the *civitates* in the Three Gauls: Belgica, Lugdunensis, and Aquitania (see Fig. 5). The purpose of the council is not fully clear: they elected a high priest of the cult of Rome and Augustus, administered funds, and emerged as a "sounding board for Gallic feeling" and a conduit of Gallic opinion to the Emperor (Drinkwater 1983:114). To be a delegate to the council or a high priest was a post of great honor and a focus of competition. This council "must be judged highly successful, satisfying hungers for status, giving the necessary religious underpinning to new political allegiances and forging a pan-Gallic aristocracy with common interests" (Wightman 1985:60).

As Wightman (1985:63) notes, "the wholesale delegation of the day-to-day administration of the Empire to the untrained leaders of the local communities did not make for efficiency." The system was run on patronage, with the consequent abuses, but it did allow the Empire to be run with extraordinarily few Roman officials.

Belgic crafts and industries under the Empire

Archaeological evidence of craft production in the cities, except for pottery, is sparse (Wightman 1985:90). The manufacture of objects and the refining of raw materials were specialties of the *vici*, the villages-cum-townships. The *vici* had their own magistrates; their inhabitants were largely engaged in craft production, especially iron making, but there is some evidence that they also market gardened or hired out as seasonal laborers to nearby villas, or agricultural estates. It is not known by what patron-client ties they may have been bound (ibid.:96).

There is remarkably little information on how crafts were organized, not only in Belgica but in Gaul as a whole. Stone quarrying and carving was a relatively new field, since Gauls were not traditionally builders in stone. Many of the early quarries were worked by Roman soldiers, others by civilian entrepreneurs from the Mediterranean who received contracts for exploitation from the army (Wightman 1985:135). Later quarries were probably privately owned, sometimes as a family business, and worked by slaves or wage laborers (ibid.:136).

Most metal ore mines probably belonged to the state, and might be worked by slaves or condemned criminals. More commonly, rights to work the mines might be leased to private entrepreneurs (*conductores*), in return for a fixed percentage of the product. Large

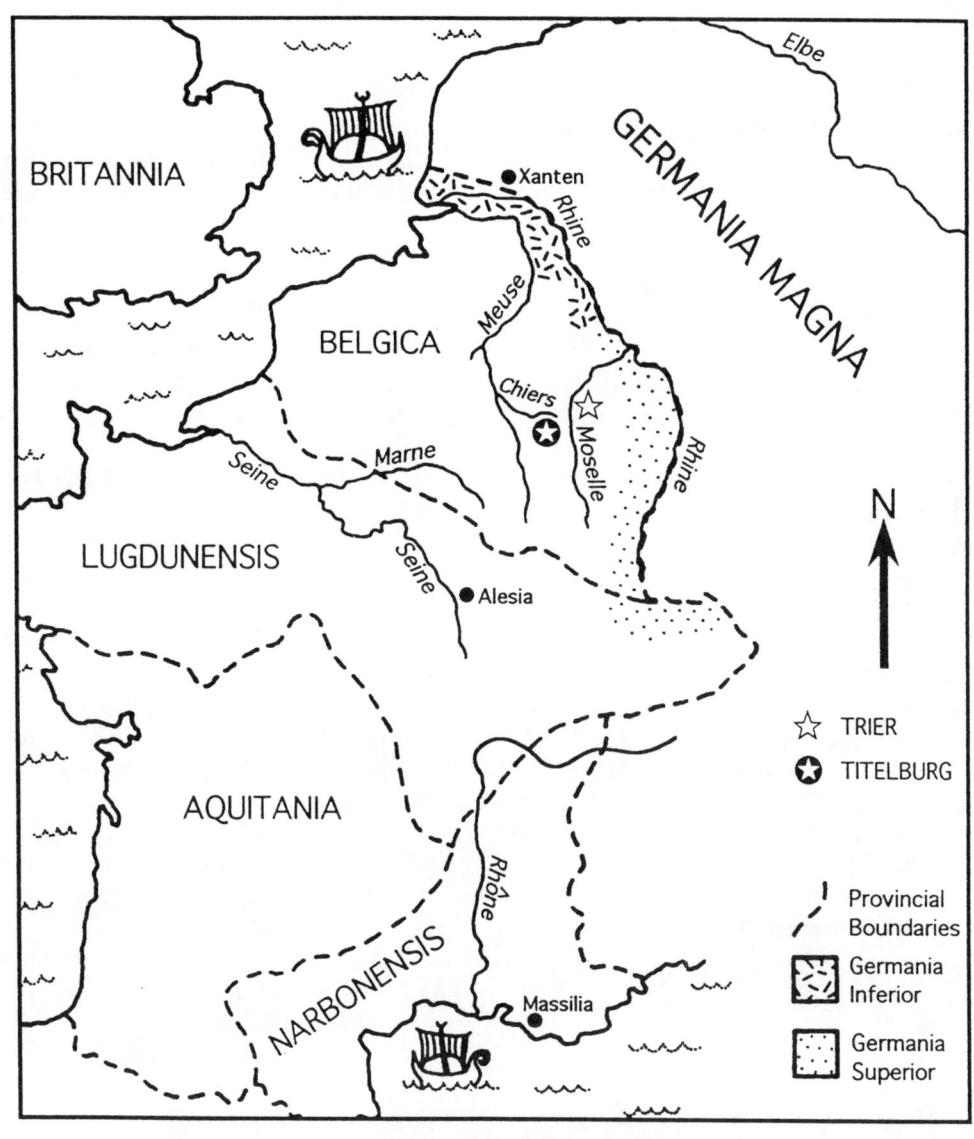

Fig. 5:
Gaul in the Augustan Roman Empire. Adapted from Drinkwater (1983:233).

slag heaps in south Belgium and Gaul attest to large-scale ironworking. Both literary evidence and the skeletons found in mines suggest that mining was a hard and sometimes fatal occupation (Pounds 1973:156).

Smaller-scale smelting and smithing production took place in *vici* and villas, especially those near the forest which provided fuel. Inscriptions suggest that most smiths were free, not slaves or bound in any way (Wightman 1985:140).

Copper mines in the Saarland south of Trier seem to have been divided into individual workings; private ownership or entrepreneurial contracting is probable. With zinc ores, imperial ownership is more likely (Wightman 1985:140); Grant (1946) suggests, based on uncited evidence, that the government held a monopoly on zinc and brass production in the first centuries B.C. and A.D. Ordinary copper- and bronzeworking took place in *vici*.

In summary, details on the Roman administration of crafts in Gaul, and especially the supervision of metal sources and metal products, is sparse at best. The army required metal, especially iron, but also copper alloys. Coin production required copper, so we may assume that the Imperial administration, small as it was in Gaul, took an interest in the metal supply. The Imperium was also interested in the supply of zinc; not only did the coinage of Augustus and subsequent emperors appear in brass but so did most military fittings of the first century A.D. We can probably assume that the supply of brass for coinage and the army came from government-owned mines whose working was contracted out to private entrepreneurs, often, at least at first, non-Gauls. Whether the production of the *vicus* workshops went to the army or supplied the local countryside, or both, is unknown.

Gallo-Roman culture

The Romans seem never to have had a consistent policy of Romanization *per se*; within a basic framework of administration, the city-states (native or Roman-founded) were left largely alone to govern themselves. Rome essentially ruled by encouraging the aristocratic Gauls to identify themselves with Roman interests (Millett 1990:37). Thus, to a large extent, the Gallo-Roman culture that developed in the conquered provinces was not so much an artifact of the Empire as an indigenous creation by the Gauls, formed of elements of their native culture, Roman culture, and new variations, acting within the constraints of the Roman administration.

We can certainly see the results of this indigenous Romanization in the archaeological record: stone houses, paved roads, imported goods, Romanized names in inscriptions. In ceramics, we see the adoption of Roman pottery styles and certain new vessel forms, implying a change in eating habits (Wightman 1985:142). Fibulae, the commonest bronze artifact in the Titelberg sample, were of Celtic origin and spread the other way to Rome (Higgins 1980:185), as did wagon technology and certain agricultural machines (King 1990:101; Wild 1976).

Gaul retained its individuality in the matter of farming methods (including the Gallic invention of a reaping machine), certain architectural types, popular art, religion and funerals, and in special industries and techniques, both because of the different environment and farming conditions in Gaul and because of rural conservatism outside the cities (Hatt 1970:176).

The invasion's effects on industrial metal production are difficult to assess. According to Collis (1984a:180) and Manning (1979), the native metal industry differed somewhat in technique (unspecified) from the Roman one and differed considerably in scale and organization, though little evidence is offered in support of this assertion. How the metallurgical industry of the Celts responded technologically to the pressures of Roman domination was one of the questions addressed in this study.

Notes

7. In Brumfiel's review of elite and utilitarian craft production in the Aztec state, she locates these part-time producers of utilitarian goods in the rural areas and the full-time independent or attached workers in the cities. In the case of the Celtic *oppida*, the existence of farmland inside the *oppida* walls may indicate a similar state of uncertainty about elite provisioning of food to clients. It also may indicate that craftworking clients had to tender food as well as craft products to their noble patrons. We simply do not know.

8. A *pagus* is a subtribe, about which little is known. It seems to have been the basic sociopolitical unit, and formed its own army and cult community (Roymans 1990:19–20). The word *pagus* is also translated as "canton" and "district."

5
THE TITELBERG AND THE ARTIFACTS FROM THE MISSOURI EXCAVATIONS

The site

The Titelberg is a large flat-topped hill located in the southwest of the Grand Duchy of Luxembourg, on a bluff 100 m above the plain of the Chiers River (Fig. 6). The large earthen rampart built in Late Iron Age times is well preserved (Fig. 7); it stands 10 m high and encloses an area of 43 ha (Metzler 1983, 1984).

The physical advantages of the site are many: it is surrounded by excellent agricultural land, it is easily defensible, and the bluff is veined by rich deposits of oölitic iron ore (Collis 1984b:173; Dunning and Evans 1986:106). Collapsed mine shafts, some worked in antiquity, can be seen on the site map (Fig. 7). The iron very likely contributed much to the site's prosperity (Metzler 1986).

Three facts make the Titelberg important for the archaeology of the Iron Age and Gallo-Roman periods: (1) it is one of the few major sites to be inhabited from the Late La Tène through the Roman period; (2) the site was abandoned around A.D. 500, and has remained uninhabited since; and (3) it is the only large Gallic *oppidum* to have had most of its excavation done in the last two decades, i.e., done to modern standards. A portion of the center of the site (Areas A, D, E on Fig. 7) has been excavated, and the results show the remains of glass making, iron smelting or smithing, and bronzeworking, especially possible coin blank (flan) casting (Rowlett and Price 1982). Roughly two thousand Gallic coins, mostly of low denominations, have been found on the Titelberg.

In addition to the iron ore under the site, copper ores could be found at no great distance; there are copper mines used at least as far back as Gallo-Roman times in the Saarland (50–60 km away) and in the Siebengebirge Range south of Bonn, approximately 150 km away (Dunning and Evans 1986).

The Titelberg was settled in the third century B.C. by people using La Tène II style artifacts; by the mid-first century B.C. (La Tène III) it was the largest settlement of the tribe of the Treveri. Judging from the large numbers of coins (Metzler 1977; Weiller 1979) and the extensive settlement, the height of the prosperity and size of the Titelberg was the third quarter of the first century B.C. Although this was immediately after the Roman conquest, the large number of Celtic coins found and the paucity of Roman imports suggest that the site was still under Celtic control. During the period of Roman control (ca. 30 B.C.–A.D. 400), the Titelberg, now off the main roads, dwindled into a prosperous *vicus*, or township (Metzler 1986), with the transfer of the Treveran capital to Trier. Though considerable glass, ceramic, and iron were produced on the Titelberg after the razing of the mint/workshop in ca. A.D. 70, the site had little administrative importance (Rowlett and Price 1982).

The archaeological investigations in the center of the site (Figs. 7 and 8) demonstrate that there was a good deal of cultural continuity between the Iron Age and the Gallo-Roman occupations. The original street plan was preserved, though the timber buildings were being constructed with stone foundations and cellars in the post-Augustan period (Metzler 1977:41). Celtic name forms continued to be used, as did the language. The Iron Age black shell-tempered pottery appears in the record until the fifth century A.D., though the percentage diminished steadily from 44% to 5% of the total ceramics (Rowlett and Price 1982).

Excavations of the State Museum of Luxembourg

Beginning in 1968, the State Museum of Luxembourg has conducted excavations both in the interior of the site and in the ramparts (see Figs. 7 and 8). In the center of the site they have discovered artifacts from the Neolithic period, and artifacts and structures from the early Roman period.

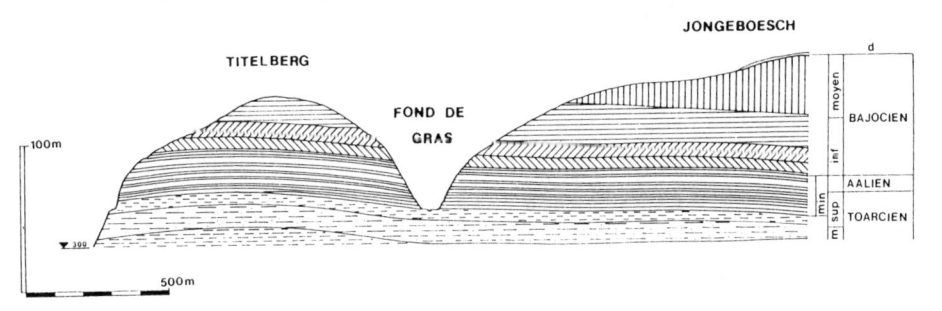

Fig. 6:
Geological profile of the Titelberg. From Krier et al. (1986:10).

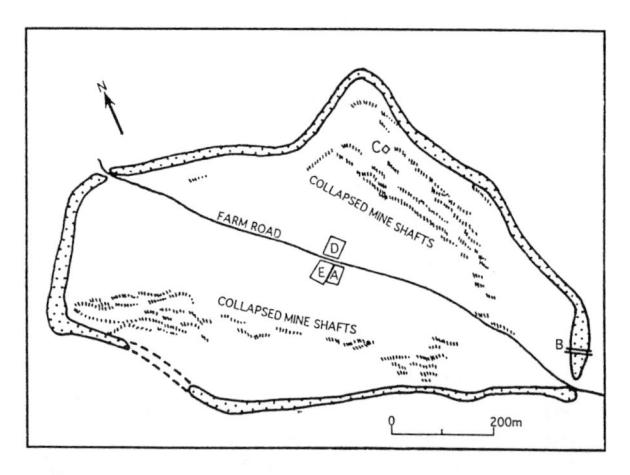

Fig. 7:
Plan of the Titelberg. The dotted area indicates the rampart. (A) indicates the University of Missouri excavation; (B–E) indicate the excavations of the Luxembourg State Museum. From Thomas et al. (1976).

The Missouri excavation

In 1972–1974, 1976–1978, and 1982, a team from the University of Missouri, Columbia, under the direction of Ralph E. Rowlett, excavated a 13 by 19 m area near the central (farm) road (Figs. 7 and 8). The area proved to be the location of a sequence of two-roomed structures (Fig. 9). Judging from the presence of clay "coin blank" molds, weights, crucibles, metal fragments, and casting debris, at least eight of the stratigraphic layers belonged to pre-Roman, early post-Conquest, and early Augustan workshops in which copper alloys were being cast and worked. In addition, many artifacts of copper-base metal were found in the soil layers above the remains of the building. We have, therefore, a series of metal artifacts (some probably manufactured off the site), covering the period ca. 125 B.C. to A.D. 300 (Rowlett et al. 1982; Thomas et al. 1975, 1976; Thomas and Rowlett 1979). This long chronological series of copper-base metal remains from one site, for an area on

Fig. 8:
Plan of the excavations carried out by the Luxembourg State Museum and the University of Missouri in 1968–1985. This shows Areas E and A on Fig. 7. From Metzler (1986:27). The area to the upper right in this plan is mapped in Fig. 9.

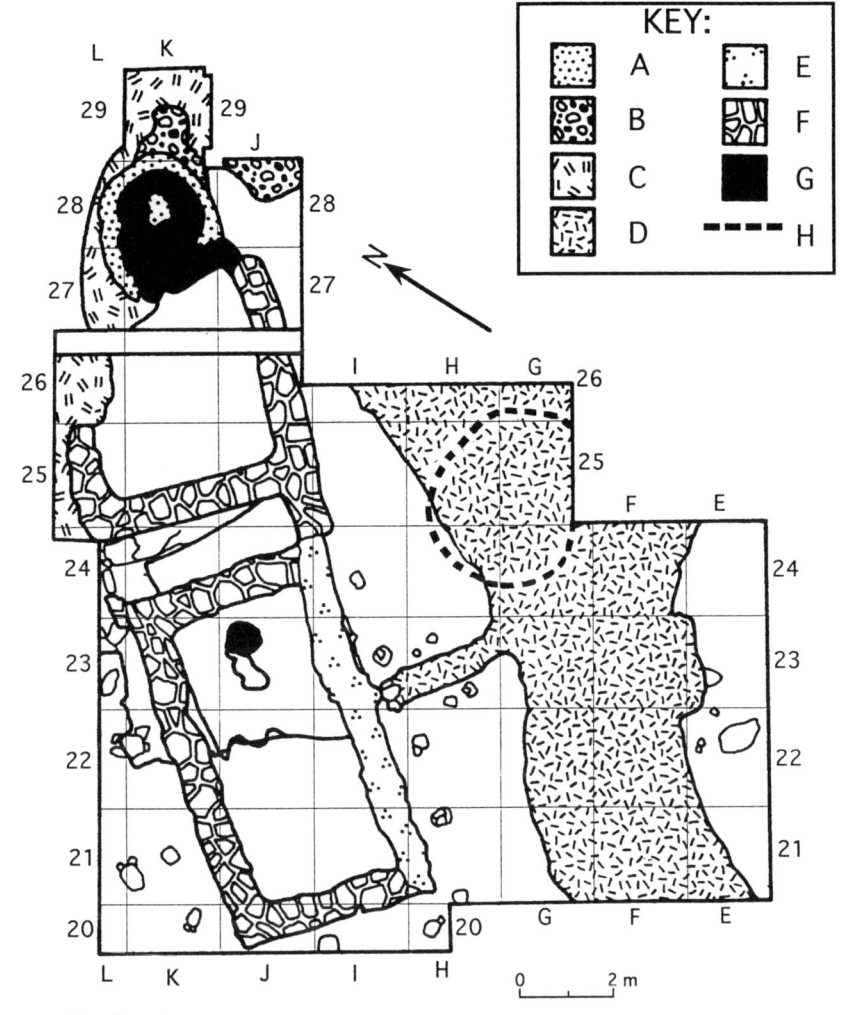

Fig. 9:
Plan of the Missouri excavation of stone foundations of workshop, associated smelters, and a side street. Adapted from Rowlett and Price (1982). Key: (A) Clay surrounding North Smelter, A.D. 4th c.; (B) stone rubble associated with North Smelter; (C) compact rubble associated with North Smelter; (D) gravel-paved side street; (E) plaster-filled robber trench; (F) foundations of Augustan workshop; (G) fireplace of Augustan workshop; (H) area of South Smelter, ca. 1st c. B.C.

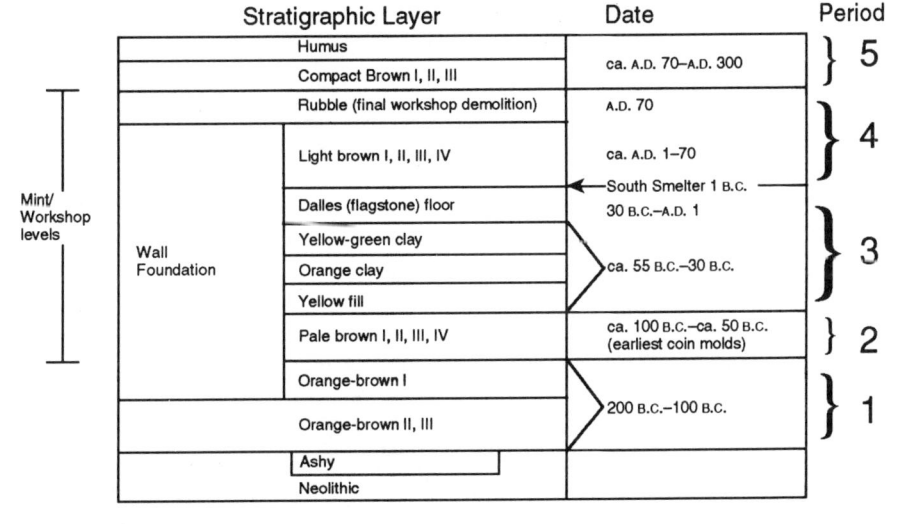

the site that was for part of that time a metal workshop, is unique in the archaeology of the European Metal Ages (Rowlett, pers. comm.).

The main features uncovered by the Missouri team are listed below in reverse chronological order (Rowlett and Price 1982; and see Figs. 9 and 10 for clarification).

- A smelter (labeled the North Smelter), A.D. fourth century, which contained a mixture of heat-damaged glass, ceramics, metal objects, and coins (see layers identified as A–C on Fig. 9).
- A non-occupation layer with small artifacts, dating from A.D. 70 to the fourth century (= Period 5 on Fig. 10).
- The foundations of an Augustan period two-room structure (4.4 by 12 m), paved with flagstones (the "Dalles floor"), which was destroyed ca. A.D. 70 (see layers and features identified as E–G on Fig. 9; = Period 4 on Fig. 10).
- A copper-base metal smelter pit (the South Smelter), used for only a brief time around A.D. 1 and refilled in one episode (see area identified as H on Fig. 9). The area was then covered with a road (see layer identifed as D on Fig. 9). The smelter fill contained residues of clay lining, slag, a tuyère tip,[9] coin blanks and molds, and the skeleton of an infant, which may represent a child sacrifice. There is a similar, though earlier, burial of a child underneath

Fig. 10:
Stratigraphy of the Titelberg workshop(s).

Stratigraphic Layer		Date	Period
	Humus	ca. A.D. 70–A.D. 300	5
	Compact Brown I, II, III		
	Rubble (final workshop demolition)	A.D. 70	4
Wall Foundation	Light brown I, II, III, IV	ca. A.D. 1–70	
		South Smelter 1 B.C.	
	Dalles (flagstone) floor	30 B.C.–A.D. 1	3
	Yellow-green clay	ca. 55 B.C.–30 B.C.	
	Orange clay		
	Yellow fill		
	Pale brown I, II, III, IV	ca. 100 B.C.–ca. 50 B.C. (earliest coin molds)	2
	Orange-brown I	200 B.C.–100 B.C.	1
	Orange-brown II, III		
	Ashy		
	Neolithic		

Mint/Workshop levels

the main roadway at Manching, a Bavarian *oppidum* (Rowlett et al. 1982).

- "A succession of at least 14 floors and 13 hearths beneath the Dalles floor" (Rowlett and Price 1982:83), which date to the period of the first century B.C. (= Periods 3 and 2 on Fig. 10). The Augustan Dalles floor building largely followed the foundations of the Iron Age buildings, which are therefore not visible on Fig. 9. The floors and structure were often renewed, but most of the foundations and all but one of the fireplaces were almost directly superimposed upon one another.
- Two Neolithic levels, the later dating to ca. 2000 B.C.

Copper-base artifacts and debris were found inside and outside the confines of the building. The complex stratigraphy of the workshop itself, in its various incarnations, covers the period from ca. 100 B.C. to ca. A.D. 70 (Fig. 10). Remains in overlying strata show that occupation of this area continued to ca. A.D. 300. Excavation of nearby portions of the site dating to this period have produced hundreds of coins and evidence of ironworking, glass smelting, and other industrial activities (Metzler 1977, 1983, 1984, 1986; Thill 1965; Weiller 1979, 1986).

The stratigraphic layers were dated by Rowlett, using fine ceramics and fibulae. For the purposes of this monograph, the layers have been divided into five chronological periods (Fig. 10).

The artifacts

The entire data set consists of 317 lots of copper-base metal material, plus 16 lots of crucible fragments and slag associated with metalworking, for a total of 333 lots (Table 3). The term "lot" is used because many of the find bags that were labeled with a single identification number have several pieces of metal inside.

These finds can be divided into several broad artifact classes:
1. Fibulae/shafts/pins (n = 50).
2. Tacks/rivets (n = 30).
3. Tools (n = 5).

Table 3. The full set of artifact lots and their chronological period

Period	Date	No. of artifacts
5	A.D. 70–300	31
4	A.D. 1–70	173
3	50–1 B.C.	91
2	100–50 B.C.	26
1	2nd century B.C.	12

4. Fittings (n = 118). "Fittings" is a catchall term, chosen because most of the remaining artifacts are of small decorative items, such as buttons, buckles, bands, bits of wire, and boss fragments, of the sort that were frequently used for horse and weapon fittings. The rest of the material in this category consists of broken objects whose function is uncertain, but which were clearly fashioned for a purpose.
5. Debris or lumps of indeterminate form (n = 114), that seem to result from casting splashes and spatters.
6. Slags, crucibles, coin-blank molds, and sherds with copper adhering (16 lots).

The entire sample set of 333 lots was used only for spatial and chronological plotting of findspots, not for metallurgical analysis. The main analysis was carried out on a subset of this data set. Some pieces were completely corroded and were discarded from the analytical program. A total of 120 artifacts, from the total available sample of 333 lots, were selected for metallography and compositional analysis. Artifacts were preferred to debris, though some debris was sampled as well. Because there were many more finds dating to the later periods than the earlier ones (Table 3), as many earlier pieces as possible were analyzed, but only a selection of later ones.

Spatial analysis of the 333 lots

The Titelberg finds were not piece-plotted, but were located in the records to the square meter. Using this information, the find squares were mapped by artifact type and chronology. The results, seen in Figs. 11–15, show that patterns of use (or at least patterns of loss) differed in the five periods. The key artifact class for tracking use-patterns is the debris class, whose deposit was likely to result from accidents in the manufacturing process.

In Period 1 (second century B.C.), the artifacts consisted mostly of debris; these were concentrated inside the area of what later became the North Room (Fig. 11). According to Rowlett (pers. comm.), the North Room was not yet built. This seems to indicate that the working area, the South Room, was cleaned regularly of its debris.

In Period 2 (100–50 B.C.), debris was again mostly confined to the northern area (Fig. 12). The discard area had expanded to the south. A number of small decorative artifacts (not debris) were found in this area. The North Room was probably built at this time.

In Period 3 (50–1 B.C.), the activity area expanded; the North Room was used as well as the South Room. Most of the debris was still concentrated in the North Room area, but 13 debris lots were found in the southern area (Fig. 13). The South Smelter dates from the end of this period; the artifacts found here are not noticeably

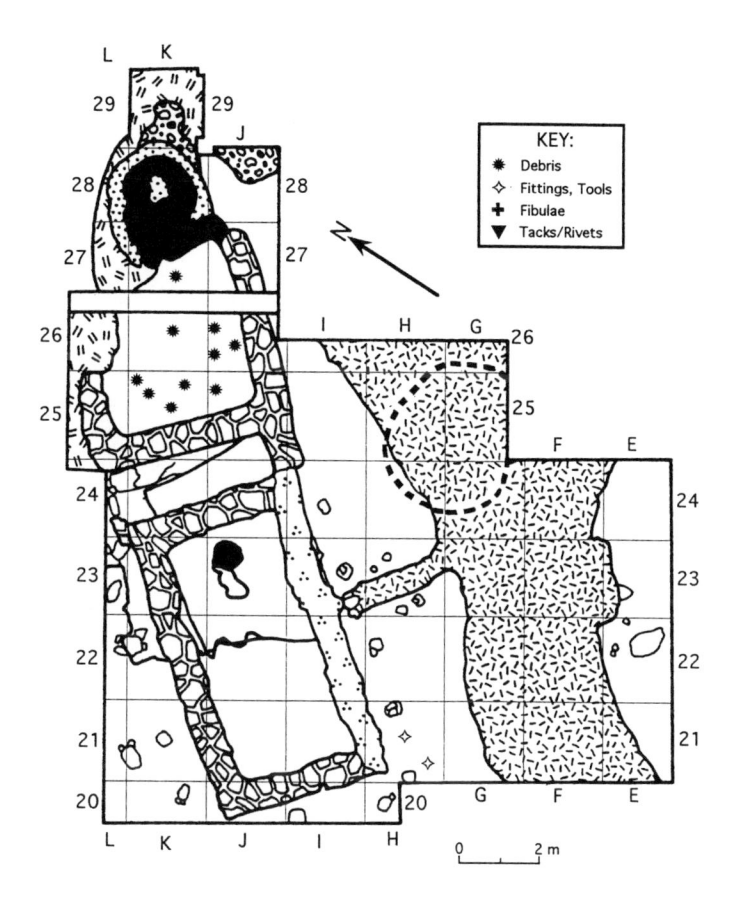

Fig. 11:
Distribution of copper-base artifacts in Period 1 (second century B.C.). See Fig. 9 for identification of features.

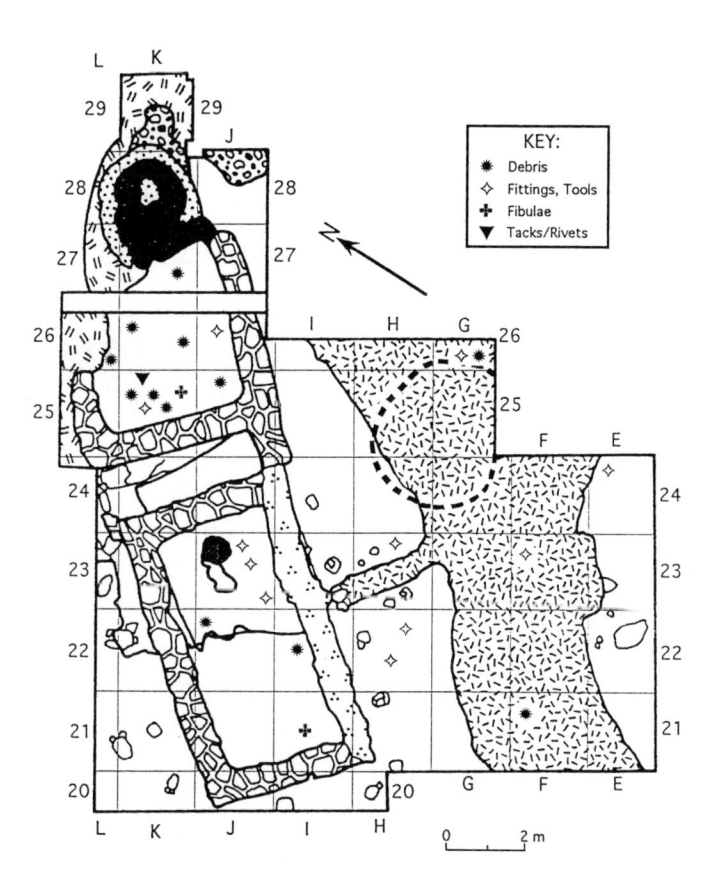

Fig. 12:
Distribution of copper-base artifacts in Period 2 (100–50 B.C.). See Fig. 9 for identification of features.

Fig. 13:
Distribution of copper-base artifacts in Period 3 (50–1 B.C.). See Fig. 9 for identification of features.

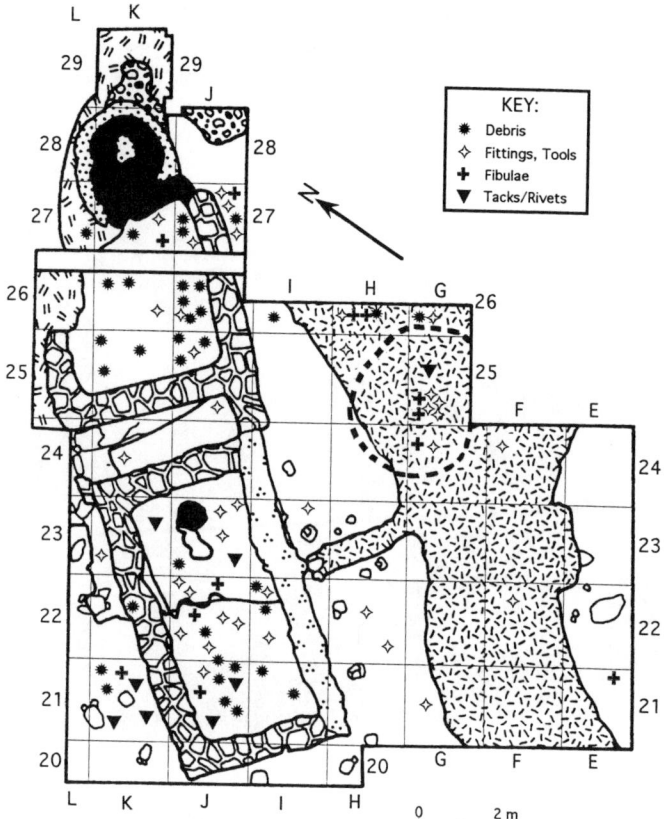

Fig. 14:
Distribution of copper-base artifacts in Period 4 (A.D. 1–70). See Fig. 9 for identification of features.

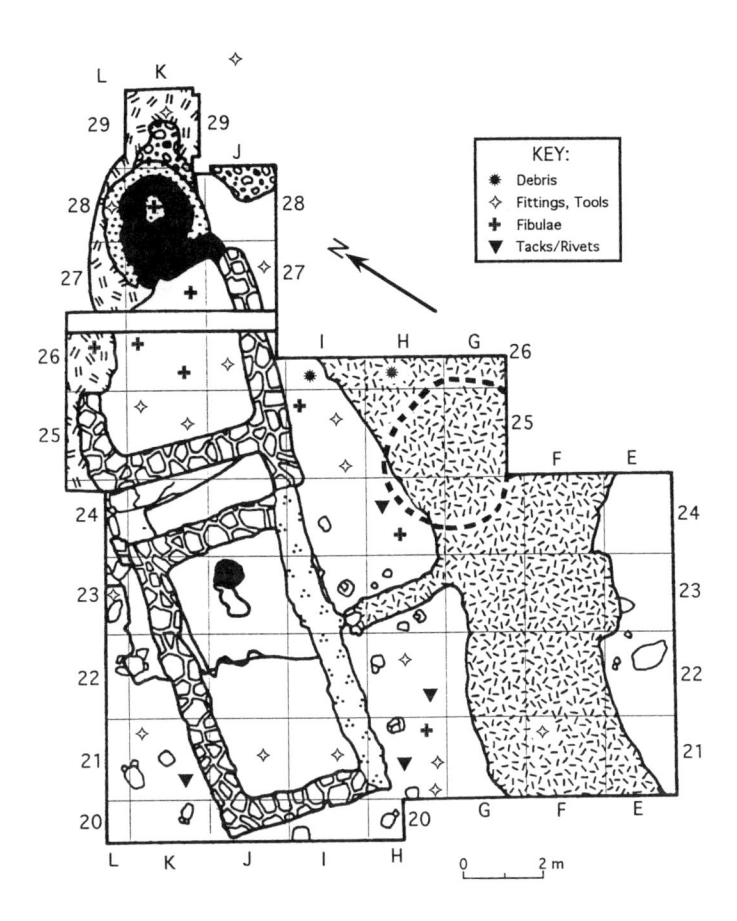

Fig. 15:
Distribution of copper-base artifacts in Period 5 (A.D. 70–300). See Fig. 9 for identification of features.

different in any way from the artifacts elsewhere, so they were put into Period 3.

A striking change is noted in Period 4 (A.D. 1–70). Almost all the remains, artifactual and debris, were found in the north half of the excavation area (Fig. 14).

In Period 5 (A.D. 70–300), we see what happened after the destruction of the building: no debris, few artifacts, and those scattered rather randomly with no clear concentrations in any area (Fig. 15).

The analyzed sample

The artifact classes for the smaller, analyzed sample are slightly different from those of the larger group (Table 4). No crucibles or slag were analyzed. In addition, the fibulae/pin/shaft group was subdivided into two groups: true fibulae and pins/shafts. While these groups often resemble each other, particularly when fragmentary, analysis shows that considerable difference in alloy composition exists between them.

As might be imagined, it is sometimes difficult to assign individual artifacts to artifact classes. This is particularly true of the small "tools" group. One tool, a needle, was unmistakable, but the others, which tended to look rather like thick awls or engraving tools, were hard-

Table 4. Frequencies of analyzed artifacts in each period, by type

Period	1 (2nd c. B.C.)	2 (100–50 B.C.)	3 (50–1 B.C.)	4 (A.D. 1–70)	5 (A.D. 70–300)	n
Artifact						
fibulae	0	2	7	14	8	31
pins/shafts	1	1	3	4	3	12
tools	0	1	3	0	1	5
fittings	1	3	15	13	7	39
tacks/rivets	0	1	9	10	2	22
debris	2	1	6	2	0	11
TOTAL	4	9	43	43	21	120

er to identify. Rowlett (pers. comm.) has suggested that they were unfinished fibulae.

Though 120 artifacts were chosen for analysis, one artifact, a tack, proved to be composed of two different alloys. These alloys were analyzed separately; hence, there are actually 121 compositional analyses.

Notes

9. Tuyères are the clay nozzles of bellow tubes; they project into the furnace and conduct the flow of air.

6
ANALYTICAL METHODOLOGY

The samples were analyzed by two laboratory techniques: metallography and compositional analysis.

Metallography

The 120 Titelberg samples were prepared by standard metallographic methods. A small piece of the artifact was removed by a diamond-bladed, oil-cooled Isomet saw. The sample was cleaned in acetone and then cold-mounted in polyester resin; the mold was cured for an hour in a 60°C oven. A few of the samples were mounted in hot-setting thermoplastic, especially if they were particularly thin-walled, since thermoplastic has better edge retention. The sample in its polyester or thermoplastic mount was then ground on 600, 430, 320, and 240 grit wet paper. The sample received a fine polish with 6 micron and 1 micron diamond paste, 0.3 micron alumina powder, and 0.06 silica suspension.

Finally, the samples were etched with a variety of etchants: aluminum hydroxide-hydrogen peroxide-water ($NH_4OH + H_2O_2$), potassium dichromate ($K_2Cr_2O_7$), or aqueous ferric chloride ($FeCl_3 + HCl + H_2O$). Slight differences in metal composition affect the performance of etchants, so the various alloys in this data set required the use of a variety of etchants to reveal their structure.

The samples were then examined under an optical metallographic microscope (available magnifications 100×, 200×, 400×, 600×), and microphotographs taken with Polaroid 4 × 5 positive-negative film, Type 55. Generous assistance in metallographic interpretation was provided by Dr. Vincent Pigott and Dr. Harry Rogers of the Museum Applied Science Center for Archaeology.

Compositional analysis

Proton-induced X-ray emission (PIXE) spectroscopy was chosen for the compositional analysis (Fleming 1985; Fleming and Swann 1993; Swann and Fleming 1986). Frequently used in medical and biological research, PIXE has been slow to be adopted for metals analysis, despite its offering an excellent combination of features suitable for archaeometallurgical study. It is nondestructive, and the machine can accommodate samples and artifacts of varying shapes and sizes. It has a high detection sensitivity, typically in the 50–150 ppm range, because using protons to excite the atoms results in far fewer extraneous X-rays (*Bremsstrahlungen*) than using X-rays or electron beams. It offers high spatial resolution, and can be operated in macro-beam (for surfaces from 4 mm^2 to 1 cm^2), to eliminate distortion due to matrix inhomogeneities, and micro-beam, to allow for specialized studies of corrosion products and slag stringers. PIXE spectrometry provides the concentrations of all the major alloying elements as well as trace elements associated with them. The actual PIXE measurements were performed by Dr. Charles Swann and Dr. Stuart Fleming at the Bartol Institute of the University of Delaware.

In sampling and grinding care was taken when possible to expose uncorroded, interior metal, so that surface corrosion and segregation effects were minimized. Twelve elements were sought by PIXE: copper (Cu), tin (Sn), lead (Pb), zinc (Zn), iron (Fe), arsenic (As), silver (Ag), antimony (Sb), sulfur (S), nickel (Ni), cobalt (Co), and chlorine (Cl). Tin, lead, and zinc are major alloying elements for copper; iron, arsenic, silver, antimony, sulfur, cobalt and nickel are important impurities; chlorine was tracked to assess the degree of corrosion and contamination by the polishing powder.[10] British Non-Ferrous Metals Standards were used; #036 for brass; #331, #361, and #391 for ternary metal; and #252, #282, and #272 for leaded bronze.

Notes
10. A chlorine percentage of 1% was considered unacceptably high, and the sample was repolished and reanalyzed.

7
RESULTS: METAL CHANGES AT THE TITELBERG

The presentation of the results is divided into two parts: the proton-induced X-ray emission spectroscopy (PIXE), or compositional, findings, and the results of the metallographic investigation.

Composition

The PIXE analyses were carried out on 120 artifacts; two separate analyses were performed on one object, a tack (#395-74), since it was composed of two noticeably different alloys. The total number of compositional analyses was therefore 121. A table of the compositional data is in the Appendix.

The major elements encountered here are copper, tin, zinc, and lead. As discussed in Chapter 3, the tin, zinc, and lead are usually added deliberately; changes in the percentages of these elements are usually the result of conscious manipulation.

The data set (n = 121) was found to comprise seven distinct types of metal (Table 5). Following Rabeison and Menu (1985), a percentage of 2% or above has been chosen as the level that indicates deliberate alloying. Percentages lower than this can be the result of accidental inclusion of an element in the ore or the flux. For example, an artifact with 97% copper and 2.5% tin was classified as a bronze; a leaded bronze would have copper, at least 2% tin, and at least 2% lead.

If the lead content is temporarily disregarded, then the result can be expressed in different form, as in Table 6. Ninety-two of the 121 artifacts (76%) fall into two groups: (1) high zinc, low tin, i.e., brass, or (2) high tin, low zinc, i.e., bronze. No artifact has more than 25% zinc, indicating that the cementation process was used to produce the brass, rather than the more difficult process of distilling pure zinc, then mixing it with copper (Hamilton et al. 1994). The cementation process requires only that copper be heated to a precise temperature in the presence of zinc ore. (See Chapter 3, pp. 16–17 for a fuller description.)

Table 5. Analyzed artifacts from the Missouri excavations at the Titelberg, by alloy type*

	N
Copper (impure)	16
Bronze	34
Brass	45
Leaded bronze	11
Leaded brass	2
Ternary alloy	5
Quaternary alloy	8
TOTAL	121

*Note: these terms are defined in Chapter 3, Table 1 and p. 14.

Table 6. Categorization of Titelberg artifacts by alloy components (adapted from Hamilton et al. 1994)

Category	N	Mean content* Zn (%)	Sn (%)
high Zn, high Sn (ternary, quaternary)	13	10.7±3.9	4.3±2.3
high Zn, low Sn (brass)	47	19.2±3.0	0.48±0.40
low Zn, high Sn (bronze)	45	0.42±0.13	10.8±4.4
low Zn, low Sn (impure copper)	16	0.40±0.05	0.18±0.39

*The appreciable scatter among the means of low levels of Zn and Sn reflects the fact that many of the data values are at, or close to, the PIXE detection limits for these elements (about 0.45% for Zn and 0.045% for Sn). The error quoted is one standard deviation.

Table 7. Changes in the Titelberg alloys through time

Period	1 (2nd c. B.C.) (n = 4)	2 (100–50 B.C.) (n = 9)	3 (50–1 B.C.) (n = 43)	4 (A.D. 1–70) (n = 43)	5 (A.D. 70–300) (n = 22)	
Alloy						TOTAL
Copper	0	0	8 (19%)	6 (14%)	2 (9%)	16
Bronze	3 (75%)	4 (44%)	17 (37%)	8 (19%)	2 (9%)	34
Leaded bronze	1 (25%)	1 (11%)	5 (12%)	4 (9%)	0	11
Brass	0	3 (33%)	12 (28%)	19 (44%)	11 (50%)	45
Leaded brass	0	0	0	1 (2%)	1 (5%)	2
Ternary	0	1 (11%)	0	2 (5%)	2 (10%)	5
Quaternary	0	0	1 (2%)	3 (7%)	4 (18%)	8
TOTAL	4	9	43	43	22	121

It is clear from this simple tabulation of alloy types that the great majority of the Titelberg artifacts fall into very distinct and deliberate alloys. Unmistakably, the smiths of the Titelberg area had control of their copper alloy production process and knew what they were producing.

Of greater interest are the changes in alloy use through time.

Alloy changes through time

The most noticeable change through time is the virtual replacement of bronze with brass and impure copper (Table 7). The first brasses appear in Period 2, or pre–50 B.C.; by Period 5, brasses and leaded brasses amount to 55% of the total. Bronze forms only 9% in Period 5. Unalloyed copper also does not receive much use until the post-Conquest period. Ternary and quaternary alloys (see Chapter 3) are concentrated in the later periods as well; though the sample size in Period 5 is half that of Period 4, the proportion of ternary/quaternary alloys in the Period 5 sample is over twice that of Period 4, and over 13 times what is in Period 3.

Also notable in Period 5 is the almost complete disappearance of lead as an alloying substance. The level of lead in the brass was always low, but Period 4 and 5 contain only one leaded brass each. Since we know, from compositional analyses of Roman statuary, that lead was common in these cast pieces (Beck et al. 1985; Caley 1970; Condamin and Boucher 1973), we must assume that the use of lead was increasingly devoted to specific artifact classes, such as figurines and statuary.

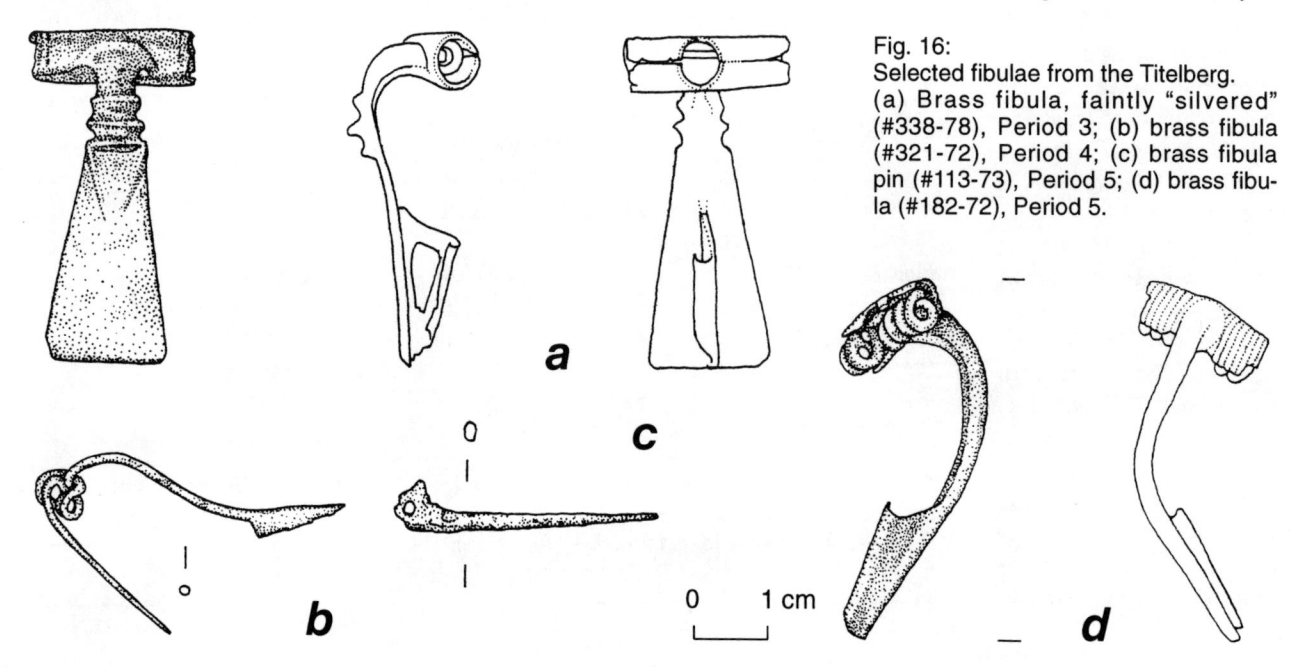

Fig. 16:
Selected fibulae from the Titelberg.
(a) Brass fibula, faintly "silvered" (#338-78), Period 3; (b) brass fibula (#321-72), Period 4; (c) brass fibula pin (#113-73), Period 5; (d) brass fibula (#182-72), Period 5.

Aside from the decline in lead, the percentages of the major elements in the brass remain the same through time.

Alloy changes in artifacts

The artifacts were divided into six groups: fibulae, shafts, fittings, tacks/rivets, tools, and debris (see Table 4 in Chapter 6).

Fibulae, or brooches

Fibulae were used to fasten tunics and cloaks; frequently they were worn in pairs connected with a chain. Fibulae were an important item of Celtic display for both men and women, particularly in the first centuries B.C. and A.D. (Fig. 16).

The most striking attribute was the heavy use of brass for fibulae and the near total absence of fibulae in bronze (Fig. 17). All but 2 of the 31 fibulae were brass or ternary/quaternary alloy. (Of the remainder, 1 was bronze and the other copper.) All the ternary alloy brooches had less than 5% tin, a relatively low amount. Without exception, the 29 brass and ternary alloy fibulae contained over 10% zinc, enough to give the characteristic golden color of brass. This use of brass for fibulae dates from Period 2 (there are no Period 1 brooches). Over half of all the brass artifacts (24 out of 45) and nearly half (5 out of 13) of the ternary and quaternary alloy artifacts were fibulae.

Shafts

Most shafts were probably hairpins, or possibly simple pins for fastening clothing (Fig. 18; for similar examples see Neal et al. 1990).

The composition of these shafts forms a striking con-

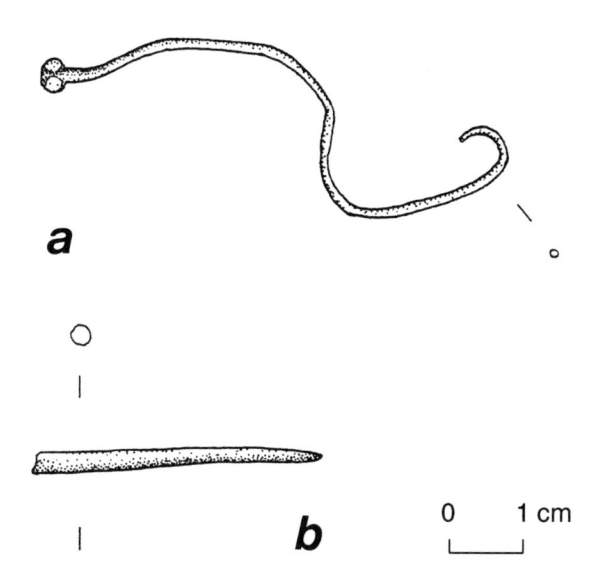

Fig. 18:
A pin and shaft from the Titelberg.
(a) Brass long pin (hairpin?) (#9-81), Period 5; (b) shaft, quaternary alloy (#318-77), Period 5.

trast to the brooches; 8 of the 12 shafts were of bronze, and 2 of the 4 remaining were of copper (see Fig. 17). Only 1 shaft was of brass, and this brass had only 11.8% zinc, one of the lowest percentages of zinc in all the brasses. Brass is softer than bronze, but there is no apparent reason why one alloy should be preferred for fibulae and another for hairpins or stickpins. The difference in color may have played a role. These hairpins/stickpins apparently occupied a different role in the area of status display or ornament.

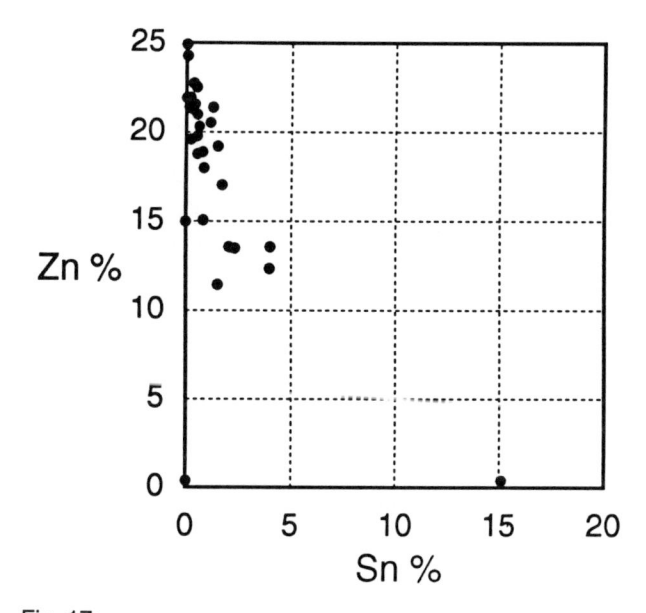

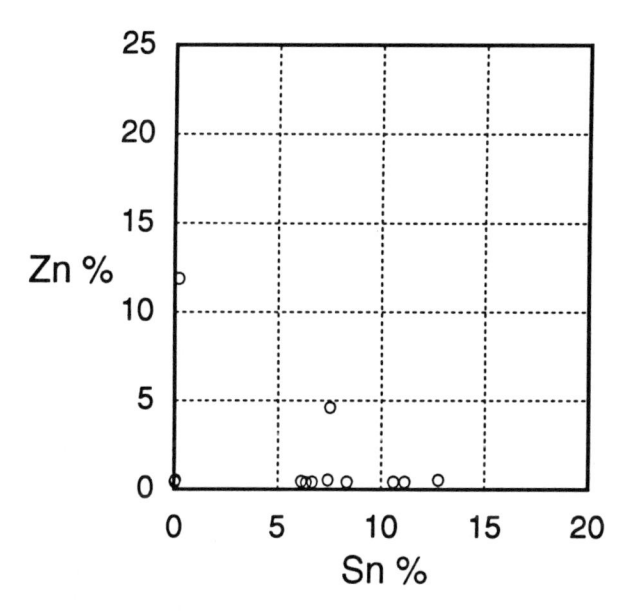

Fig. 17:
Scattergrams, zinc vs. tin, for Titelberg fibulae (solid circles) and pins and pin shafts (open circles). From Hamilton et al. 1994.

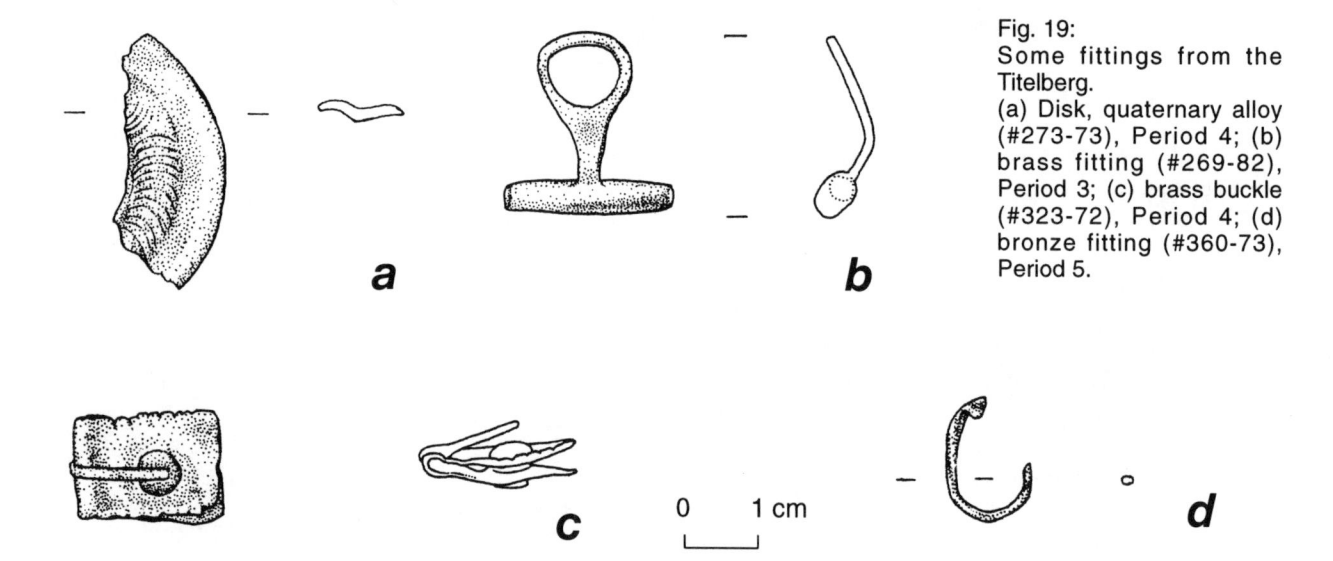

Fig. 19:
Some fittings from the Titelberg.
(a) Disk, quaternary alloy (#273-73), Period 4; (b) brass fitting (#269-82), Period 3; (c) brass buckle (#323-72), Period 4; (d) bronze fitting (#360-73), Period 5.

Fittings

This is a general group of small ornamental objects, probably mostly from horse harnesses, military kits, or vessel decorations (Fig. 19).

This items in this category are approximately evenly divided between bronze (19) and brass (15), with 2 of copper and 2 of quaternary alloy. The bronze tends to be concentrated in the earlier periods (1–3); the brass and quaternary alloys in the later periods (4–5).

Surveying the range of shapes in this heterogeneous category, there seem to be few instances of consistent association of alloy with artifact shape, save that the more complex and finely finished artifacts in the earlier periods tended to be of leaded bronze. The lead in the alloy would dictate the use of casting, since it is difficult to forge leaded copper alloys. In the later periods the more complex artifacts were of brass.

Tacks/rivets

The tack-like objects are very variable in size (Fig. 20). Their exact function is unknown. They range from a thick rivet 5 cm long to tiny spikes a centimeter long and 2 mm wide. They again showed a consistent association of alloy; 10 of the 23 tacks were of copper, and 7 of these were tiny spikes. Only 8 tacks/rivets were of brass or ternary/quaternary alloy. The reason behind the preference for copper in tacks is unknown.

Tools

This is an ambiguous category, comprising 1 bronze needle (#125-74) and 4 items that could be engravers or scoring tools (Fig. 21). Rowlett's suggestion that they are unfinished fibula backs is supported by their compositions: 3 of them were brass and 1 was copper. If they were indeed tools, they would have been intended for

Fig. 20:
Some tacks/rivets from the Titelberg.
(a) Copper tack/rivet (#727-78), Period 3; (b) copper tack/rivet (#10-78), Period 3; (c) bimetallic tack/rivet, ternary alloy/copper (#395-74), Period 5; (d) tack/rivet, quaternary alloy (#708-77b), Period 4.

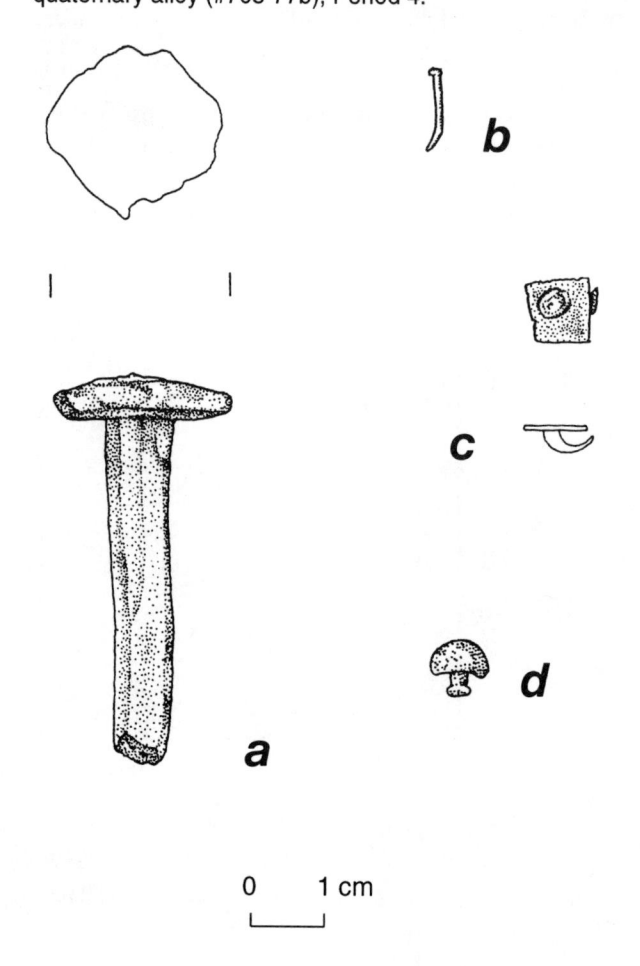

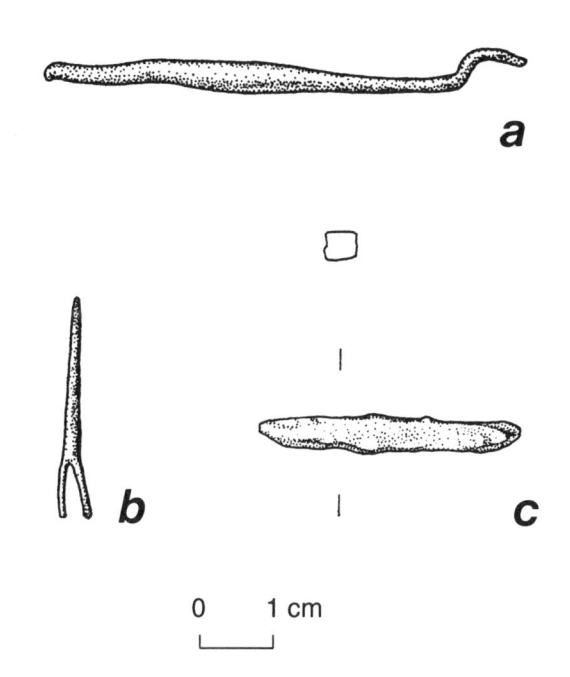

Fig. 21:
Some tools (?) from the Titelberg.
(a) Brass tool (?) (#508-73a.1 and a.2), Period 3; (b) bronze needle (#125-74), Period 2; (c) copper tool or unfinished fibula back (#523-78), Period 3.

engraving, which would require a harder metal than copper or brass. Bronze is suitable, but by this time such tools were usually made of iron, as seen at the *oppidum* of Manching (Jacobi 1974:pls. 5–8).

Debris

The group of debris is made up of amorphous lumps, presumably accidental spills or waste from manufacturing.

All 11 debris pieces were of bronze. Since it is likely that these are the remains of castings and manufacturing processes—a supposition borne out by the disappearance of debris after ca. A.D. 70, when the mint/workshop was torn down—we may assume that the metal actually processed in this succession of buildings was bronze. The average percentage of tin was fairly high (mean = 12.77%), with percentages going up to 19%. It is possible, given the large quantity (over 800 pieces) of ceramics that Rowlett and Price (1982) identified as "coin blank molds," that this debris results from the making and casting of bronze or *potin* coins.

The only good collection of Gallic and British native cast and struck copper-base coin analyses is that of Northover (1992). The coins came from British excavations and British and Continental museum collections. There are significant differences in the composition of struck and cast coins. The Gallic and Helvetic cast coins have tin contents of 5–40%; the British cast coins have tin percentages between 13% and 37%. The cast coins of the Durotriges, a tribe in southern Britain, have even less tin: 5–19%. Lead levels are variable but the coins usually have several percent of lead. The struck pieces, on the other hand, have very low tin (0.03–6%, most on the lower side) and only trace amounts of lead. They would, in fact, be better called struck coppers (Northover 1992).

The Titelberg debris pieces, with 10–16% tin and 0.9–6% lead, fall comfortably within the lower range of the cast potin coins, but clearly outside the range of the struck coins. (*Potin* is usually defined as a high-tin alloy used for some Gallic coins, but clearly in some circumstances the tin was not very high.) One would be inclined to say, on the basis of the analyses, that potin coins were being cast here were it not for the presence of the coin blank molds. Some of these molds have been analyzed using X-ray fluorescence and neutron activation analysis; they seem to have been used to cast "bronze" and gold (Thomas et al. 1976). (The composition of the "bronze" is not stated; the word tends to be used in the archaeological literature to refer to any sort of copper or copper alloy.) These coin blank molds are usually assumed to have been used for the production of coin blanks of specific weight, the blanks later to be struck by dies (Tournaire et al. 1982). Thus the Titelberg material is contradictory: the debris composition suggests that either bronze coins were being cast or some other leaded tin bronze article was being manufactured; the molds found suggest that copper-base and gold coins were being struck. Only inspection and analyses of the numerous Celtic native coins found at the Titelberg—inspection hitherto not permitted—will solve this problem.

Trace elements

The minor impurities or trace elements considered were iron (Fe), arsenic (As), silver (Ag), antimony (Sb), and nickel (Ni). They are considered apart from copper, lead, tin, and zinc because it is unlikely that these trace elements were added deliberately. They result from elements already in the ores and are altered by smelting practices. While efforts to use impurity patterns to trace ore proveniences have been fruitless (see p. 12), the range of variation of impurities can provide some clues to the homogeneity of the ore sources. Usually the percentages of these trace elements were well under 1%, though four artifacts had over 1% antimony, one had 5% silver, and one had 1.1% arsenic.

Analysis of the distribution of these minor elements through time has revealed some complex patterns.

Iron

Iron percentages increase somewhat throughout the five periods (Figs. 22–24), though never exceeding

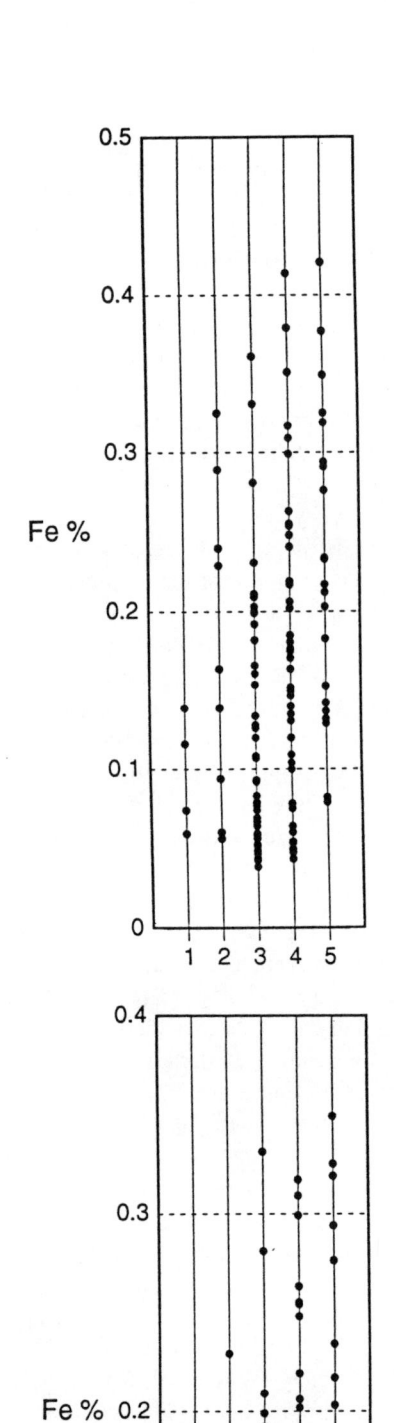

Fig. 22: Percentages of iron vs. period for all alloys.

Fig. 23: Percentages of iron vs. period in bronze.

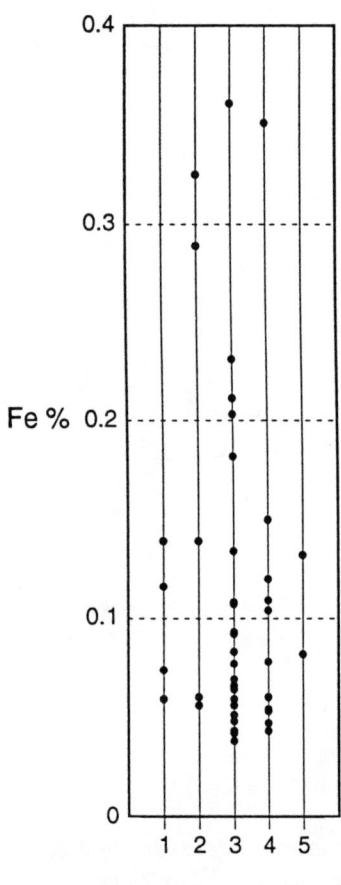

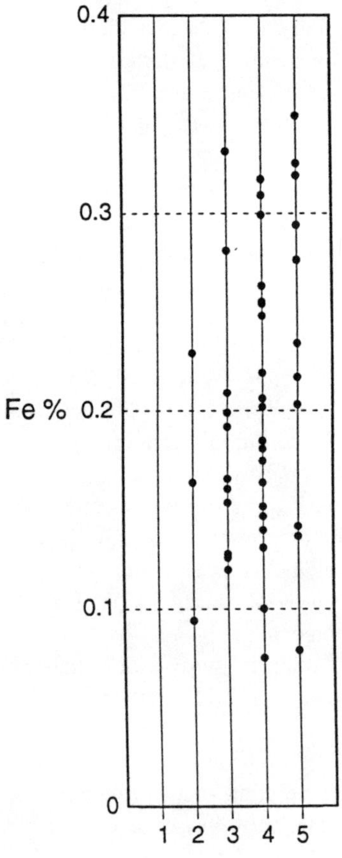

Fig. 24: Percentages of iron vs. period in brass.

Fig. 25: Percentages of arsenic vs. period for all alloys.

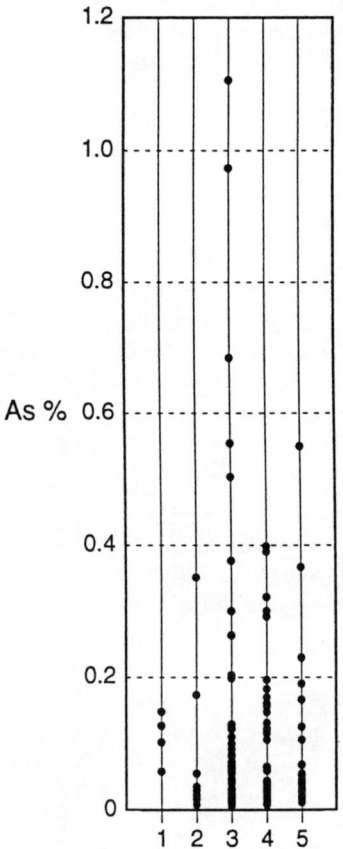

Table 8. Trace element means (without outliers)

	Element			
	As (%)	Ag (%)	Sb (%)	Ni (%)
Bronze	0.173	0.219	0.293	0.159
Copper	0.137	0.218	0.203	0.241
Tern./Quat.	0.161	0.102	0.104	0.099
Brass	0.048	0.055	0.053	0.100

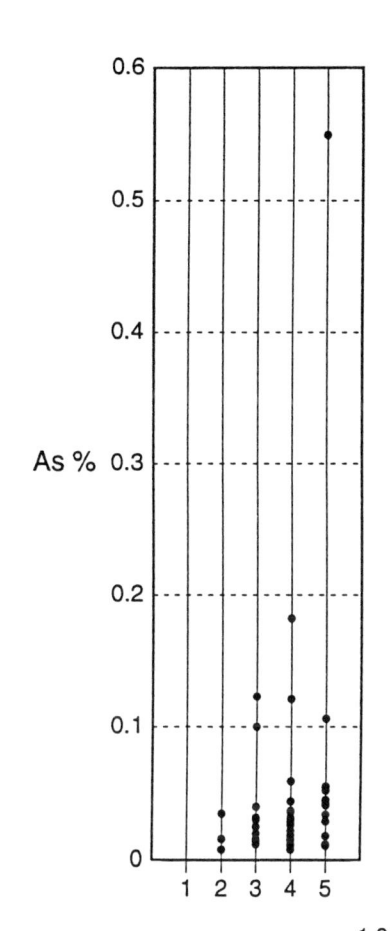

Fig. 26: Percentages of arsenic vs. period in brass.

0.421%. Rabeison and Menu (1985) discovered in their analysis of artifacts from Alesia that iron levels tend to rise with zinc levels. The iron may be in the zinc ore (Fig. 24).

Arsenic

Arsenic varies in complex ways through the sample. The level of arsenic is consistently low in brass (mean = 0.048) and rises in copper (mean = 0.137), ternary/quaternary alloys (0.161), and bronze/leaded bronze (0.173) (Table 8 and Figs. 25 and 26). Looking at As percentages through time, a spike of high As is clear in Period 3 for both bronze and copper alloys. The same is reflected, of course, in artifacts made predominantly from one alloy. The mostly bronze shafts have more As in Period 3, and fibulae, predominantly brass, contain only a low percentage of As in all periods.

Further analysis suggests that the arsenic is a byproduct of lead additions. Eight of the 11 artifacts with higher As had lead percentages of 2% or more (as contrasted with 21 high-lead samples out of 121 for the data set as a whole). These high-lead percentages range from 5.637% to 25.67%. This suggests that a new source of lead was used in Period 3, one that was distinguished by higher As content.

Silver

Again, complexities are revealed. Figure 27 shows a clear pattern; silver percentages were less than 0.5% in Periods 1, 2, and 5, and climbed as high as 0.913% (and, in one unusual artifact [#338-78], to 5.04%) in Periods 3 and 4. What is notable, after dividing the sample into brass and non-brass (Fig. 28), is that the Ag level in the non-brass vastly exceeds the percentages in the brass (Table 8) except for the aforementioned #338-78 (this is a fibula with a silvery sheen; the silver may be a surface application, though no evidence of this was found in the PIXE and electron microprobe analyses). With the exception of this fibula, the percentages of Ag are nowhere high enough to have been a deliberate addition. It appears that different copper ores were used for brass and non-brass.

Fig. 27: Percentages of silver vs. period for all alloys, without outlier.

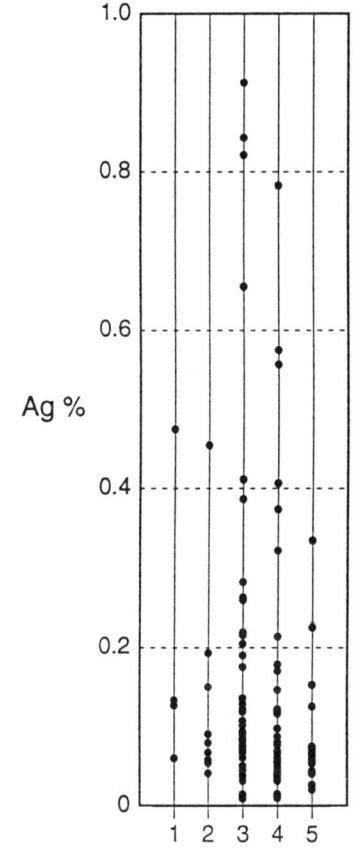

Antimony

A similar pattern is noted in antimony. Again, with a single outlier (#528-72, with 6.24% Sb) disregarded, the mean antimony percentage in bronze was 0.293%, and for brass it was 0.053% (Table 8 and Figs. 29 and 30). One can note heightened Sb percentages in Periods 2–4, with a sharp decline in Period 5. The antimony percentages in the brass were far lower, with no decline from the already low content in Period 5.

Nickel

A similar though less pronounced pattern is seen with nickel (Table 8 and Figs. 31 and 32). Bronze and copper had higher means of nickel than did brass or ternary/quaternary metal, but both brass and bronze from Periods 3 and 4 had higher percentages of nickel.

Implications of the compositional results

The analyses of the major elements show that bronze was, from the middle of the first century B.C. to the middle of the first century A.D., largely replaced by cementation brass and impure copper. This is particularly true for fibulae; as early as Period 2 (100–50 B.C.), the single fibula was of brass. Fittings, too, were increasingly made of brass, though this begins later than with fibulae; it is only in Period 4 that the bare majority of the fit-

Fig. 28: Percentages of silver vs. period for brass, without outlier.

Fig. 29: Percentages of antimony vs. period for all alloys, without outlier.

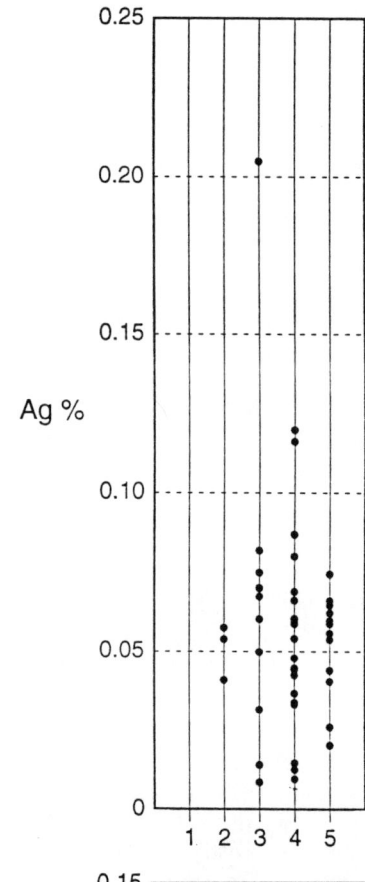

Fig. 30: Percentages of antimony vs. period in brass.

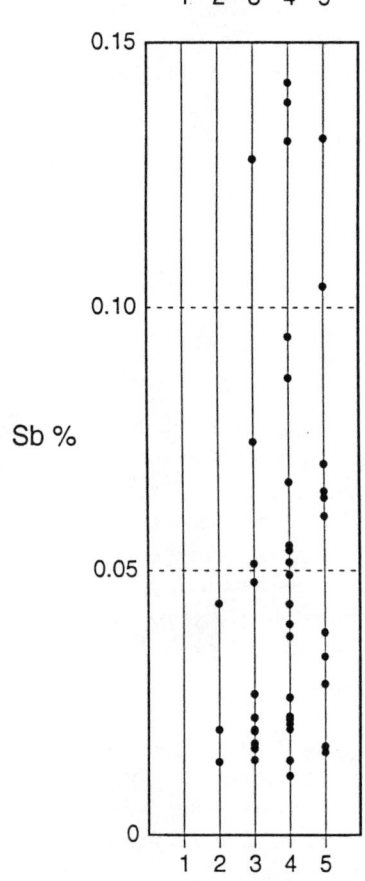

tings were of brass. Tacks/rivets started out being made of bronze in Periods 1 and 2, but by Period 3 most were of copper. Shafts/pins show a pattern that is the reverse of that for fibulae; 75% of them were of bronze, with only one brass shaft appearing, in Period 5. Debris, the final group considered here (tools are not considered because of the ambiguous artifact attribution), were all of bronze, indicating that the metal actually being manipulated in the workshop was a relatively high-tin bronze. If the four ambiguous "tools" are indeed unfinished fibulae, then it is unlikely that the metal for them was produced in the building, though they might have been produced elsewhere on the Titelberg and been sent to the workshop for forging. None of them had the high-lead percentages that would preclude hot or cold working.

We also note the increase in mixed alloys, or ternary and quaternary alloys; these increase from 11% (one artifact) in Period 2 to 28% (six artifacts) of the Period 5 sample set. These were almost certainly the result of the opportunistic smelting together of bronze, copper, and brass scrap pieces, though Craddock (1978) believes that this was a deliberate manufacture of a general-purpose alloy.

Using the trace elements, one finds a similar clear separation in the ore used to make the brass metal from the copper and bronze ore sources. This holds true for Periods 2–4, but it is difficult to determine for Period 5 because of the paucity of bronze and copper from that period.

In addition, one can detect ore use variation within the bronze-copper group. The percentages of all the trace elements are consistently low in Periods 1 and 5; variation begins to increase in Period 2 and greatly increases in 3 and 4. With arsenic one can pinpoint this further; its greatest range of values by far occurs in Period 3. All this implies that a homogeneous ore source was used in the second century B.C., and probably for most of the first half of the first century B.C. From the Conquest until ca. A.D. 70 a greater range of ores was used. It is possible that the Treveran craftworkers had access to new sources of ore with their incorporation into the Roman polity, or, given the prosperity of the Treverans in the first few decades after the Conquest, economic demands may have forced a diversification of ore sources. We do not know; there is no other information on expansion of ore sources. Whatever the explanation, the ores used seem to have become more homogeneous after A.D. 70 and the destruction of the mint/workshop. (It is possible, however, that the change is more a reflection of the decline in bronze and copper use as compared to the more homogeneous brass rather than any true restriction on ore sources.)

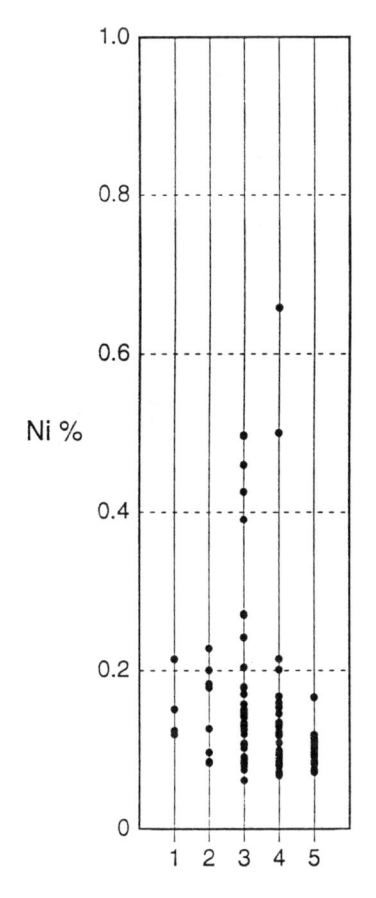

Fig. 31: Percentages of nickel vs. period for all alloys.

Fig. 32: Percentages of nickel vs. period in brass.

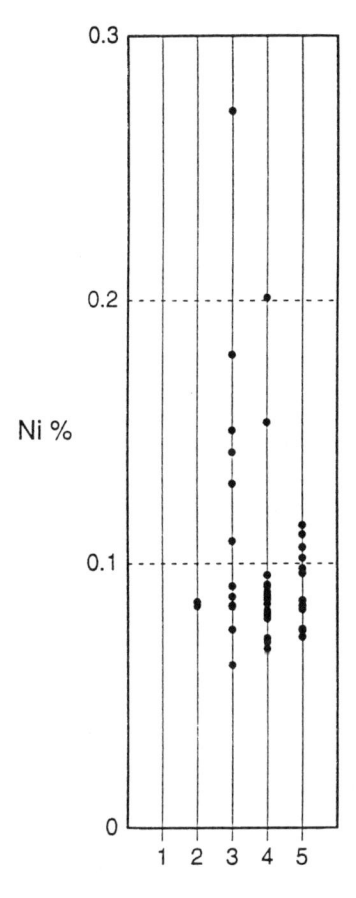

In summary, the PIXE analyses demonstrate clear patterns of alloy change through time, as well as clear preferences for the use of certain alloys for certain artifact classes. In addition, they show the difference between the metal sources of brass and non-brass, as well as use of a variety of ore sources during the period of initial Republican and Imperial consolidation. After A.D. 70, the metalworkers appear to have returned to the homogeneity of ore sources seen in the Late Iron Age.

Metallography

The metallographic structures revealed in the course of the research and recorded in 121 photomicrographs fit fairly comfortably into ten major groups. In the pages below, each group is described and illustrated by at least one photomicrograph.

The presentation of metallographic structures begins with dendritic structures, though recrystallized granular structures predominate in this study group. Of course, sometimes there are no clear dividing lines even between dendrites and recrystallized grains, as the section on "ghost" dendrites will show.

Group 1

This group consists of unworked dendrites, i.e., unworked castings that have not been heated long enough for full homogenization. Figure 33, a lump of debris from Period 3 with 19% Sn, shows a very clear dendritic structure, with an unusually large piece of redeposited copper in the middle. Most of the debris pieces, as one might expect, show unaltered dendrites, but so do several fittings as well. By and large, though, these North Gallic smiths did not favor unworked castings. Even some of these clear dendritic structures show the beginning of recrystallization or homogenization after annealing. This is seen in Fig. 34, a fitting composed of a quaternary alloy dating to Period 5: 62% Cu, 25% Pb, 8% Zn, and 4% Sn.

Group 2

The structures in Group 2 are related to those of Group 1; these are samples that show clear granular structures as well as "ghost" dendrites left behind after incomplete annealing. Figure 35 shows these ghost dendrites in a small leaded brass fitting (17% Zn, 10% Pb) of the late first century A.D.

Ghost dendrites can also be left behind by corrosion that preferentially attacks one phase in a dendritic pattern, as illustrated in Fig. 36, a ternary alloy (14% Zn, 2% Sn) fibula of the second century A.D.

Group 3

The structures in the third group are columnar grains, a variant of dendrites. They result from casting

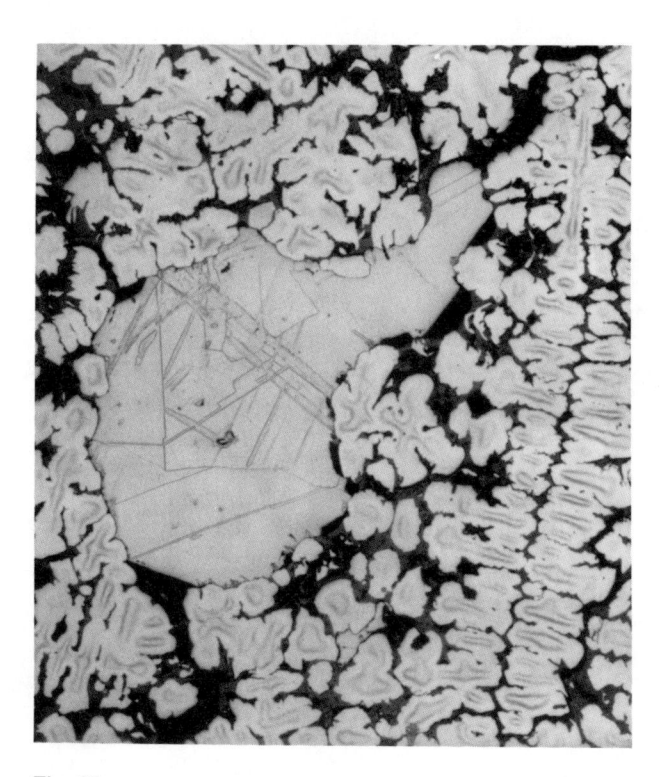

Fig. 33:
Bronze debris (#125-82). Dendritic cast structure with considerable corrosion and large area of recrystallized copper. Group 1. 100×. NH$_4$OH + H$_2$O$_2$.

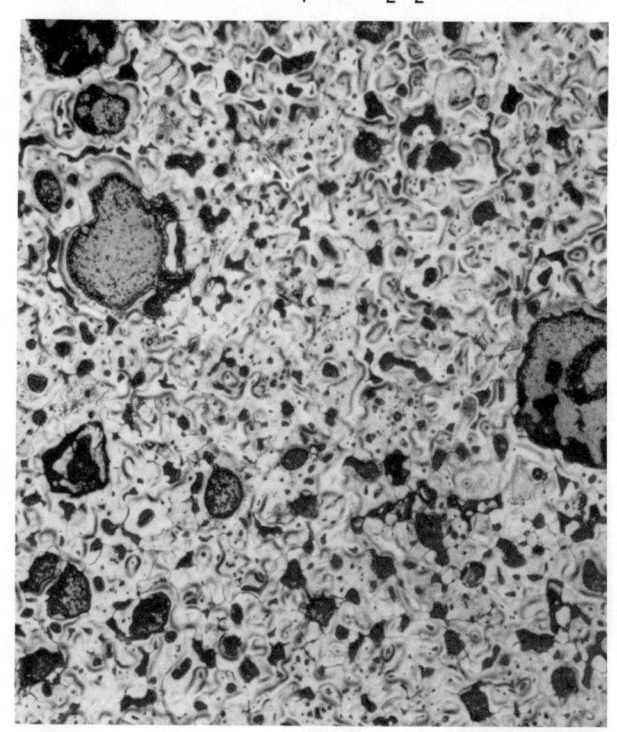

Fig. 34:
Fitting, quaternary alloy with 25% lead (#525-76). Very clear dendritic structure, but grain boundaries are forming between arms of dendrites. Large lead globules are visible. Cast and annealed. Group 1. 100×. K$_2$Cr$_2$O$_7$.

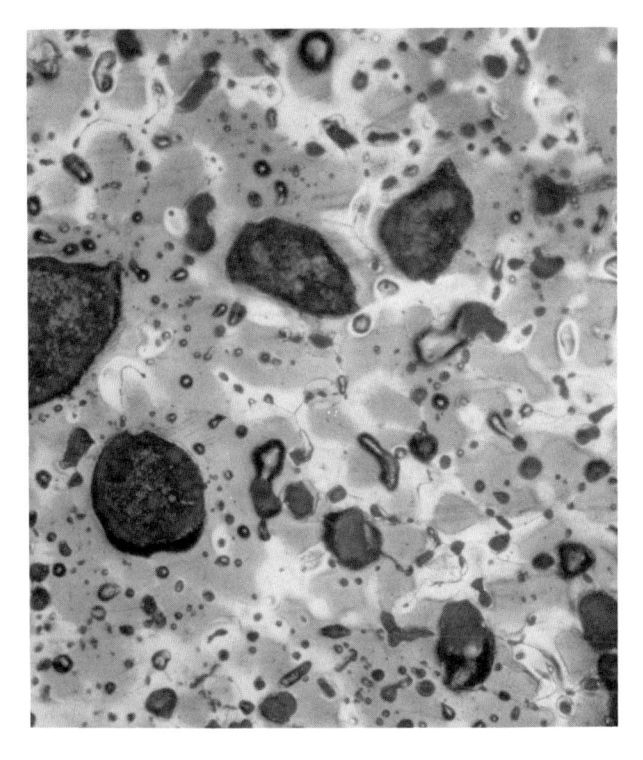

Fig. 35:
Leaded brass fitting (#133-73). Cast, followed by slow cooling or reheating. Some recrystallization with a few strain lines. Group 2. 200×. $NH_4OH + H_2O_2$.

with erratic cooling. There are only two samples that display this rare structure, both of impure copper. The figure (37) shown is of a fitting of ca. A.D. 1–70.

Group 4

The fourth group is characterized by plain, fully homogenized, unworked grains without visible distortion or annealing twins. Again, firm boundaries are sometimes hard to draw, for frequently in a sample there will be a small patch with annealing twins that indicate prior working and heating, with the larger part of the polished surface showing no sign of alteration. Often the intergranular boundaries will be extensively corroded. Figure 38 is a piece of second century B.C. debris that was fully recrystallized. It is an 11% Sn bronze with 1.4% Pb.

Group 5

The structures in the fifth group consist of equiaxed grains with annealing twins. This is by far the largest group and indicates the most popular and consistent method of treating copper-base metal. Clearly, repeated cycles of working and annealing were the order of the day. As is discussed below, the grains tend to be remarkably small, from ASTM #6 to 10 or even smaller. Also frequent are long strings of inclusions/slag, which also

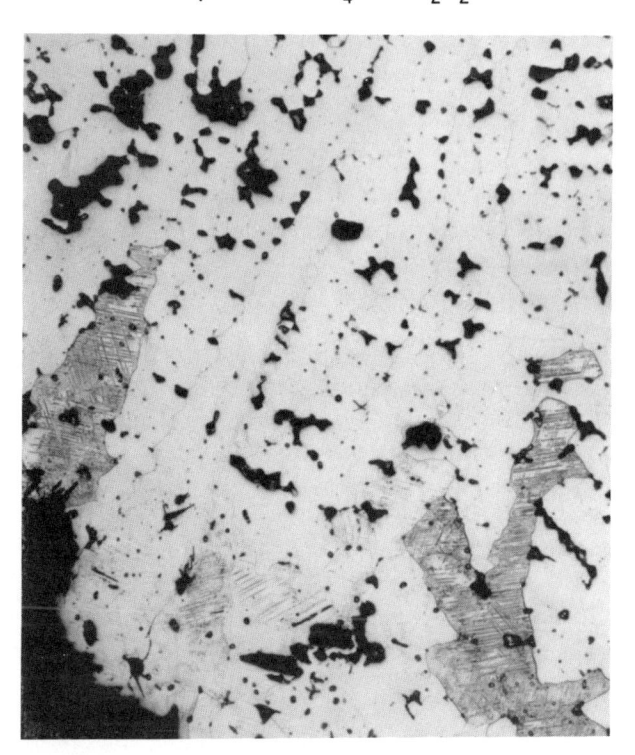

Fig. 36:
Fibula, ternary alloy (#446-76). Cast and reheated to beginning of recrystallization. Dendrites are seen clearly in corrosion pattern. Group 2. 100×. $K_2Cr_2O_7$.

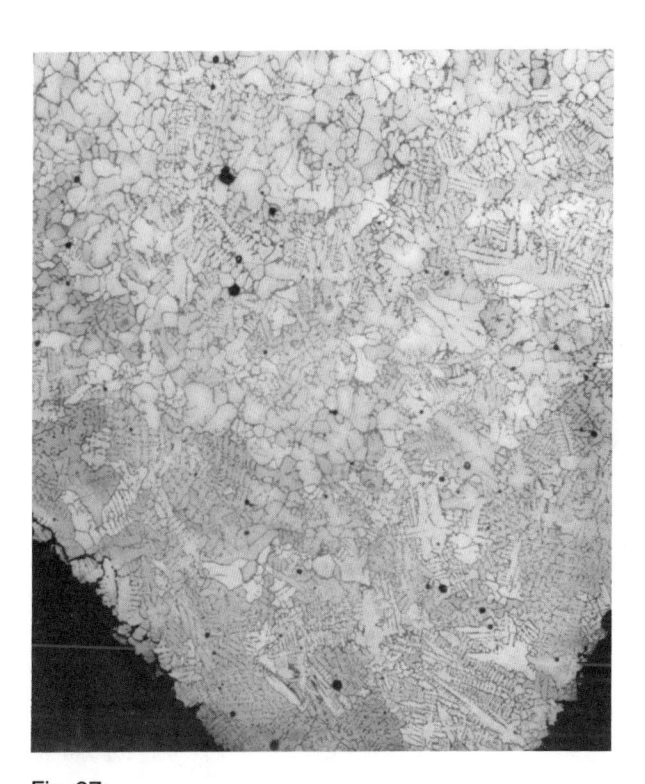

Fig. 37:
Small copper fitting (#137-77). Cast and erratically cooled. Columnar grain growth in several areas. Group 3. 100×. $K_2Cr_2O_7$.

Fig. 38:
Bronze debris with 1.4% lead (#455-82). Worked and annealed; heavy grain boundary corrosion. Group 4. 100×. $K_2Cr_2O_7$.

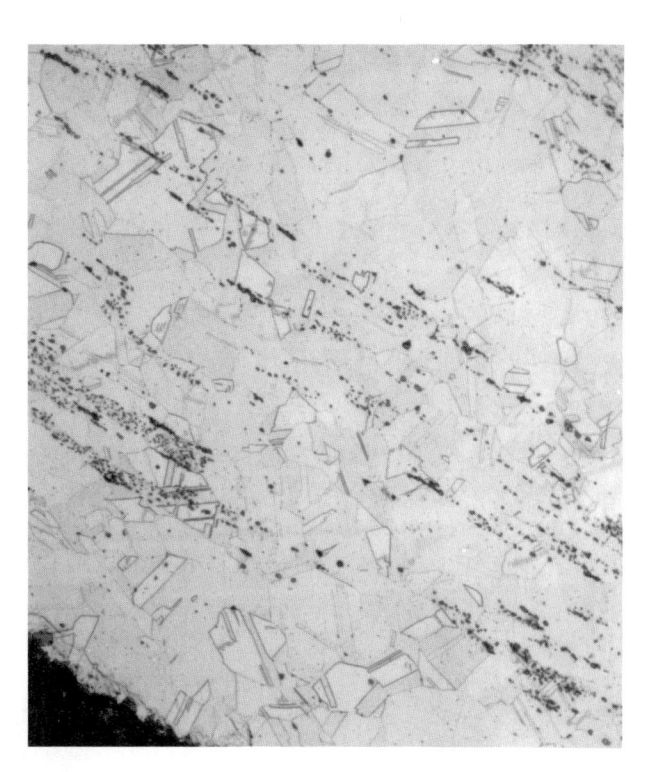

Fig. 39:
Brass pin (#9-81). Recrystallized grains with annealing twins; numerous slag stringers parallel to length. Group 5. 200×. $NH_4OH + H_2O_2$.

Fig. 40:
Brass fibula (#207-72). Recrystallized grains with annealing twins. Group 5. 200×. $NH_4OH + H_2O_2$.

reveal the practice of repeated hammering. For an example of this, see Fig. 39, a long hairpin of 12% Zn brass, dated to the third century A.D. Figure 40, a first century A.D. brass fibula, illustrates annealing twins in relatively large grains. Figure 41, a low-tin bronze of Period 5, illustrates annealing twins in the more frequently seen smaller grains.

Group 6

The sixth group consists of tiny grains in metal that has been so severely cold-worked that the structure looks flowed or rippled. This pattern results from a series of workings and annealings, followed by a final, very severe working. The most extreme example can be seen in Fig. 42, a fitting (a ring or earring) of 9% Sn bronze dating to ca. 100–50 B.C.

Group 7

This group is comprised of samples that have been given the same treatment as in Group 6, but with a less severe final working. The grains—or dendrites, since the distortion is so severe that often no distinction can be made between worked homogeneous grains and worked dendrites—are merely heavily flattened and not rippled. The sample shown here (Fig. 43), a relatively mild

Fig. 41:
Bimetallic rivet (#395-74). Tiny recrystallized grains with twins. Worked, annealed, some reworking? Numerous small irregular inclusions. Group 5.100×. NH$_4$OH + H$_2$O$_2$.

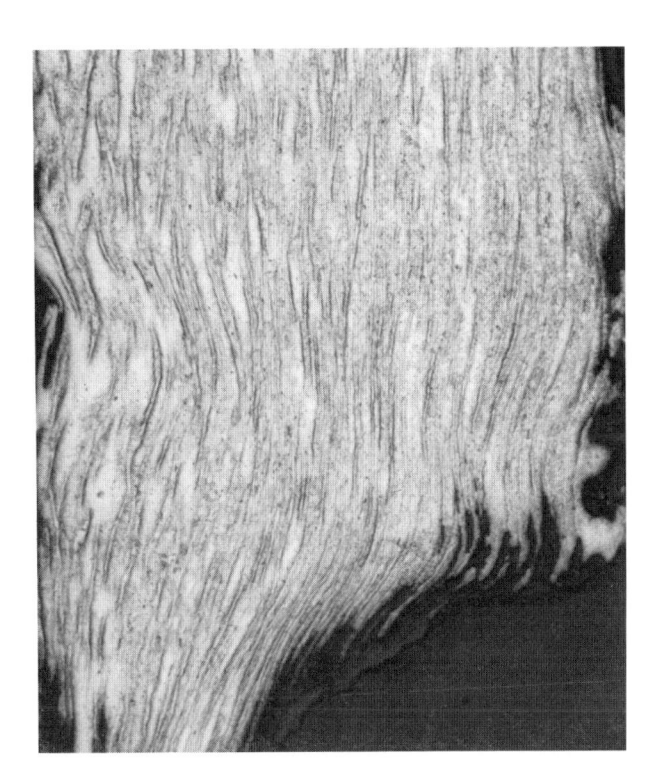

Fig. 42:
Bronze fitting (#226-74). Heavily worked, annealed, very heavily worked. Extremely tiny grains in a flowed matrix. Group 6. 100×. K$_2$Cr$_2$O$_7$.

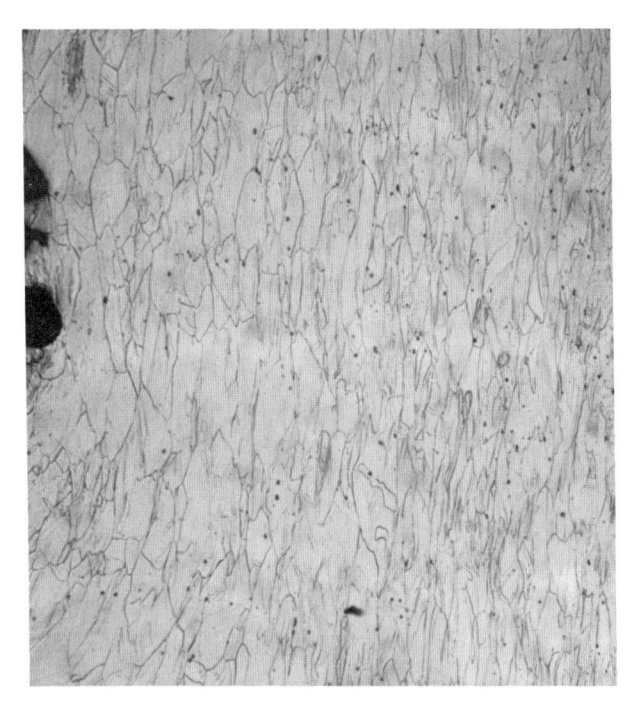

Fig. 43:
Head of copper rivet (#727-78a). Worked, annealed, reworked. Grain flattening. Group 7. 100×. NH$_4$OH + H$_2$O$_2$.

example, is from the edge of the head of a very large rivet; the shank of the rivet shows equiaxed grains with straight annealing twins. The figure shows bent and deformed twins.

Groups 8 and 9

These two groups are variants of one another. They are related to the previous two groups, the difference being that they have received a less severe final cold working. The grains are less distorted, but the working has produced numerous thin (Group 8) or thick (Group 9) strain lines or stress marks in the grains. Corrosion tends to follow these strain lines, as can be seen in Fig. 44 (a fibula of ca. A.D. 1–70, composed of a ternary alloy of 14% Zn and 4% Sn). Much thicker strain lines and bent twins can be seen in Fig. 45, a brass (21% Zn, 1% Sn) fibula of the same date.

Group 10

The last group consists of the eccentrics, those structures that are anomalous and difficult to understand. Two examples will suffice. Figure 46 is of a small artifact that resembles a animal paw; the alloy was leaded bronze, with extraordinarily high tin: 23%. There was also 6% Pb and 1% Sb. The structure is undecipherable,

Fig. 44:
Fibula, ternary alloy (#912-77). Worked, annealed?, worked. Heavy strain markings, though no visible grain flattening. Group 8. 200×. $K_2Cr_2O_7$.

Fig. 45:
Brass fibula (#289-73). Recrystallized grains with annealing twins and heavy strain lines. Group 9. 400×.

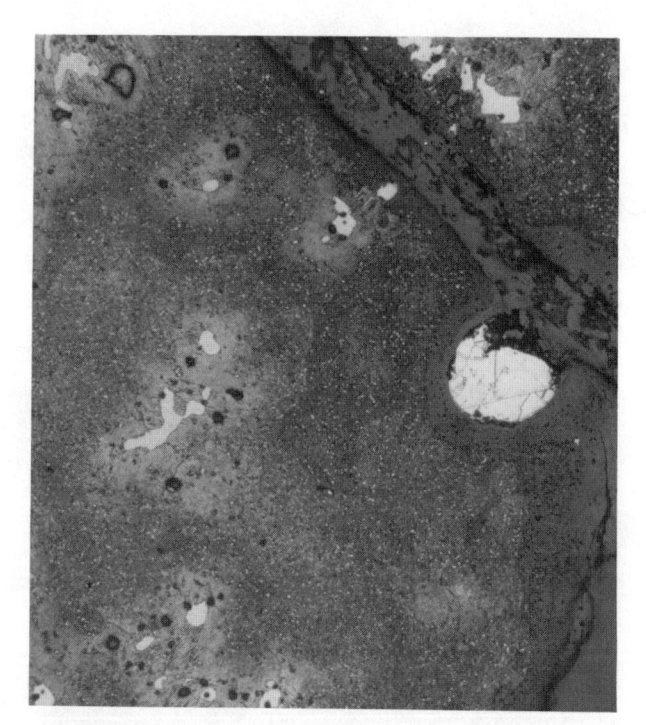

Fig. 46:
Fitting of leaded bronze with 23% tin and 1% antimony (#346-82). Cast. Unusual dendritic structure. Group 10. 400×. $K_2Cr_2O_7$.

with 2 to 3 different precipitates. A portion of the Sn has gone into the Pb. The process that produced this structure is unclear, but there seems to have been little working involved. Figure 47 is of a fitting, a small ring dated to ca. 1 B.C. This is a heavily leaded bronze alloy, with 26% Pb and 6% Sn. The photomicrograph shows at least three phases, one an alpha+delta eutectoid. Copper has been redeposited in the Pb globules. The artifact was cast.

These eccentrics illustrate that alloy production was by no means uniform. Most of these eccentric structures results from bizarre quantities of alloying elements rather than from bizarre metallurgical treatments. We have, at this late date, no way of knowing whether these strange alloys resulted from accident, incompetence, or an experiment.

Summary of metallographic findings

Metallographic examination revealed a diversity of manufacturing techniques, but with some strong trends. Leaded bronze, leaded brass, and quaternary (Cu+Sn+Zn+Pb) pieces had dendritic structures, showing that, unsurprisingly, they were nearly all cast, with little if any post-casting treatment. Five of the bronze pieces also retained a dendritic structure. Most of the

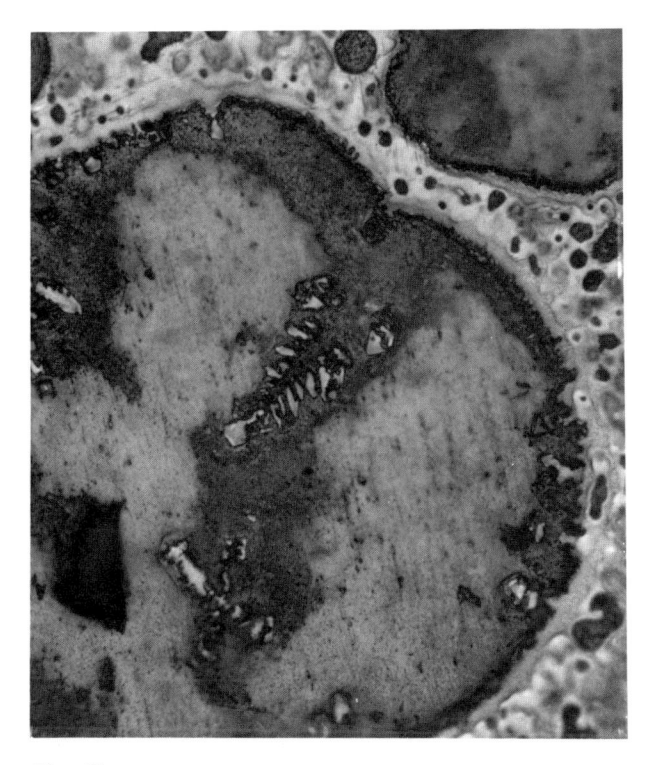

Fig. 47:
Fitting, bronze with 26% lead (#192-78). Cast structure, multiphase, with large lead globules. Group 10. 400×. $K_2Cr_2O_7$.

artifacts, however, are characterized by remarkably tiny, annealed grains with signs of extensive and repeated cycles of working. Small grains can be produced by a variety of metallurgical conditions: intensive cold working, annealing at a low temperature or for a short time, and the presence of impurities or alloying elements. One clue that the grain size is probably due to repeated cycles of cold working and annealing is the presence of long, parallel strings of slag and inclusions (see Fig. 36 above for an example). While the grains of the metal itself recrystallize and assume an equiaxed shape, the deformed slag inclusions do not; they remain flattened. The longer the strings, the more severe the cold working that produced them (Michael Notis, pers. comm.). Many of the Titelberg artifacts contain these long stringers of slag or other inclusions.

As a general rule, the Titelberg artifacts show signs of heavy working, and this changes little through time. In the absence of comparable metallographic analysis of Italian Roman artifacts it is impossible to know if this is a sign of the retention of traditional northern Gallic habits or a metallurgical tradition common to Europeans on both sides of the Alps.

The copper artifacts show the most consistent treatment; five of the seven show considerable distortion, a sign that heavy cold working was the final step taken. Most of these copper pieces were "spikes," or small, slender tacks/rivets often made by folding flat sheet.

The ternary alloys—alloys of copper, tin, and zinc—received the most diverse treatment: one had dendrites; some had small grains, some had very small grains with inclusions, and some showed large grains. These unleaded alloys were not produced out of ignorance or carelessness, since the smiths cold-worked far more of them than the quaternary alloys that contained lead, but the degree of cold working received seems to have been quite variable.

There seem to have been no gross changes in metallurgical treatment from Periods 1 through 5. All the periods contain examples of pieces left in a cast state, as well as pieces heavily worked and annealed.

The variety of production methods can be seen when the data set is broken down by artifact type. The tacks/rivets consistently show recrystallized grains with no dendrites, and tend to show signs of extensive working. Fibulae usually show tiny twinned grains, with six exceptions; four pieces exhibit casting dendrites (two of them ghost dendrites) and two show large untwinned grains, the result of reheating cast metal.

The fittings show more diversity than the fibulae: nine were left cast, two were hammered after recrystallization and left unannealed, seven recrystallized after casting but were never cold-worked, and the rest exhibited small twinned grains.

The most interesting group is the debris. One would expect this group to be composed of unaltered dendrites, the result of rapid cooling after being spilled from a crucible or mold, and indeed 8 of the 11 are. But the remaining 3 show fairly large twinned grains. It is possible under certain circumstances to obtain annealing twins without prior working; rapid cooling can produce enough stress so that twins are formed (Notis, pers. comm.). The presence of three presumed debris pieces that show recrystallization and twins, however, is difficult to explain; it does seem as though these pieces were worked and reheated.

8

CONCLUSIONS: TECHNOLOGY AND SOCIETY AT THE TITELBERG AND IN GAUL

In the first chapter, I indicated that the principal objective of this monograph was to examine in detail one aspect of "Romanization" at one site in what became the Roman province of *Gallia Belgica*. I proposed to do this by examining, through compositional analysis and metallography, the specifics of 400 years of copper-base metal production.

This objective has been reached and the results make it clear that the metalworking does not show a straight "Romanization." Rather, as was stated in Chapter 4, an indigenous creation was formed by the Gauls of elements of their native culture, of Roman culture, innovations from outside Europe, and new variations.

The previous chapter presented the results of the analyses. Now the implications of these findings need to be addressed. The results throw light on several aspects of Gallic culture and culture change in the years between 100 B.C. and A.D. 300.

The first use of these new data can be to establish fresh facts about the particular culture history of this area. It was demonstrated that there was indeed a regular pattern of changes in metallurgical practices over the period of incorporation into the Roman Empire. Bronze and leaded bronze dominated in the earlier periods, then were in part replaced by brass and copper. The brass and the bronze seemed to come from different ore sources, and different ore sources were exploited for the non-brass artifacts in the hundred years after the Conquest. Alloys were tailored to artifacts: most fibulae were made of brass, hairpins and shafts of bronze, tacks of copper.

It is commonly assumed that brass was introduced into transalpine Europe by the Romans after the Conquest (Tylecote 1976). The first widespread use of brass in Italy was in the 23 B.C. coinage of Augustus.[11] The Titelberg evidence, however, shows 3 brass artifacts (from a sample of 9) dating to 100–50 B.C. (see Fig. 48 and Table 7 above), and 12 more (from a sample of 43) dating to 50–1 B.C. All have zinc percentages ranging from 11% to 23% (mean = 19%), strongly suggesting that the metal was produced by the cementation method. Before the Augustan coinage of 23 B.C., the only place in the Roman and Mediterranean world where the widespread use of cementation brass was attested was in certain parts of western Asia Minor. This Titelberg evidence, then, shows not only that the Gauls were using brass earlier than has been assumed heretofore, but that they were using it earlier than the Romans.

Though Celtic mercenary activity in the eastern Mediterranean had largely ceased by the beginning of the second century B.C. (Griffith 1935), Celtic tribes still lived in Galatia in Asia Minor and may have provided a conduit for brass in the form of coins. It is also possible that the knowledge of the cementation process spread from Galatia to Gaul via these tribes or possible Gallic travelers. Whatever the source—raiding or trading—the brass was brought back home and ended up being melted down into ornaments. It should be remembered that brass, or orichalcum, was highly valued before the invention of the cementation process made it relatively inexpensive to produce (see above, Chapter 3), and it may be that this reputation clung to the golden-looking alloy in the first century B.C. After 23 B.C., of course, the Celtic metalsmiths need only have melted down Roman brass coins to obtain their metal, or carry out their own cementation process. In the first century A.D., Roman military fittings would have provided another source.

In any case, the earliest brasses in the Treveran land did not remain coins, if indeed that is how they started. In Period 2 (100–50 B.C.), the three artifacts of brass were

1. a long rod of bent metal that is almost certainly the loosened spring and body of a one piece fibula (#720-73),
2. a 2 cm long flattened hook (#703-73), and
3. an 8 cm long band or strap (#1053-73) that Rowlett (pers. comm.) has identified as being stylistically German.

These are all personal ornaments or military/horse fittings (Fig. 48).

The high levels of zinc and the infrequency of ternary/quaternary alloys in the earlier periods suggest that the brass for the artifacts was either being made afresh, or had only been melted down once, for when brass is remelted it loses some 10% of its zinc (Bayley 1990:21). No decline in zinc levels through time is seen, but since material from A.D. 70 to A.D. 300 is lumped together into Period 5, the analyses would not reflect the documented decline in the zinc content of Roman brass

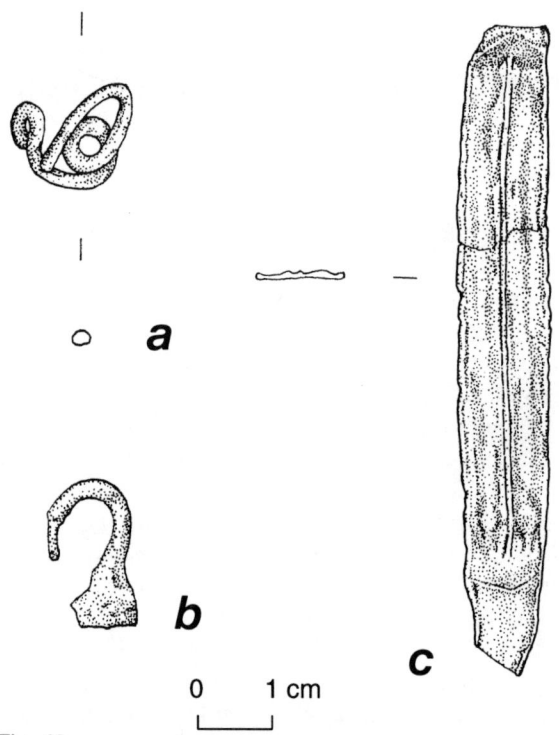

Fig. 48:
The three brass artifacts that dated to before the invasion of Caesar: (a) a loosened fibula spring, #720-73, (b) a fitting, #703-73, and (c) a fitting, #1053-73 (see text for discussion).

coinage after Claudius (Caley 1964; Riederer 1974a). Nonetheless, the rising percentage of ternary/quaternary alloys, to 27% in Period 5, suggests that bronze and brass scrap were increasingly being melted together. Only 9% of the artifacts in Period 5 were made of bronze and leaded bronze, from 100% in Period 1, and the average tin percentage in the Period 5 artifacts is the lowest of all the periods.

Regional differences

Evidence produced in other studies, notably Beck and colleagues 1985, Carter 1971, Condamin and Boucher 1973, and Rabeison and Menu 1985, demonstrates that workshops and even mints in Gaul had their own distinct recipes for alloy production. From the excavations in Alesia (Beck et al. 1985), it is known that metal ingots circulated not as pre-made alloys, but as pure copper, allowing smiths to manufacture articles according to their own formulae. Condamin and Boucher (1973) discovered that workshops in central Gaul and the Rhineland used more zinc than other areas. Beck and colleagues (1985) discovered that, in contrast to Roman practice, nearly half of the Gallo-Roman figurines from France they analyzed had added zinc. Woimant and Hurtel (1989) described a very Gallic-looking statuette of a god-warrior as being of 20% zinc brass.

Remelted scrap in these areas retained specific lead isotope patterns and trace impurity patterns, which implies that the scrap, too, circulated within a relatively small area. As Beck and colleagues conclude (1985), this strongly suggests that while ideas and general technology spread widely, the actual movement of smiths and products was limited.

Ore sources

The impurity patterns presented in the last chapter show that the brass metal contained a much narrower range of trace element percentages than the non-brass metal did, which suggests that the source of the copper was different and much more homogeneous. Carter and Buttrey (1977) have similar findings for the copper and orichalcum coins of Augustus and Tiberius; the copper used in the orichalcum coins had distinctly lower nickel contents than the metal used for copper coins. This suggests that this specialization of source was a phenomenon widespread in the western Empire, at least in the early years.

The Titelberg data set shows an increase in mixed alloys in the post–A.D. 70 period, presumably caused by opportunistic melting together of scrap bronze and brass; the British data show the same (Bayley 1990). Analysis of post-Claudian coins (no mint specified) shows the progressive debasement and loss of zinc in copper-base coinage. The care taken to manufacture fresh brass and bronze declines, and while alloys are still tailored to some artifact classes (Riederer 1983, 1984), there appears to be a growing indifference to the distinction between brass and bronze. While the workshop excavated at the Titelberg cannot support this, since it was destroyed ca. A.D. 70, nor the Alesian workshops, since they seem to date to the first century A.D., one can offer a hypothesis that the care taken in earlier periods to separate the copper intended for brass and the copper intended for other alloys vanished, as did the separate workshops that specialized in brasswork. This can only be tested by chronologically controlled excavations of post–A.D. 70 workshops.

Workshops

The cementation process for manufacturing brass was difficult and specialized, and the analysis of the debris at the Titelberg excavation shows no indication that brass was being manufactured or worked in that building. All the analyzed debris shows that this particular smithy building seems to have been devoted to manufacturing items of bronze. So there seems to have been spatial differentiation in bronze- and brassworking sites at the Titelberg. This is not a unique situation; a similar spatial specialization is found at Alesia (Rabeison and Menu 1985), where two workshops contained only brass debris.

Belgic Gaul and the anthropology of technology

The social value of metals

All artifacts possess a certain social value in addition to their functional value, and all artifacts convey meaning as one of their functions. They can convey status (high or low), gender, age and wealth distinctions, religious and ethnic affiliations; they can be valued as sacred or despised as unclean, restricted to nobles or fit only for slaves, suitable for gifts or an insult if given. Though even the most utilitarian objects convey cultural information, it can be hypothesized that the less immediately utilitarian an object or material is, the more cultural information it encodes, simply because virtually its entire reason for existence is to signal data about the user or owner.

After the use of iron became widespread in temperate Europe, the copper-base metals largely assumed this kind of informational function. Certainly artifacts of copper and its alloys could and did perform utilitarian functions: fibulae held cloaks and tunics together, hairpins fastened coils of hair, bronze vessels held liquids. All these functions could have been performed by iron or other materials, and in fact some fibulae were made of iron. Given the lack of any functional necessity to use copper-base materials, another reason must be sought for the decision to employ them. The copper-base metals were used primarily for display and ornaments, and eventually for coinage. The use of a metal for coinage is a particularly good indicator of its display status, for the value of a metal coin is usually in inverse proportion to the degree that the metal is used in ordinary tools. What is displayed would be prestige, authority, or the power of the state or leader to monopolize or control valuables.

It is clear as well from the result of the Titelberg sample analysis that the people who made and used these copper-base artifacts valued the various alloys of copper rather differently. Brass was used earlier in Belgic Gaul than it was in Italy, and it was soon largely allocated to fibulae. Bronze was retained for pins and shafts, which may have been hairpins or may have been a simpler form of clothes fastener than fibulae. Rivets, whether large or small, were usually of impure copper. After the turn of the millennium, the use of brass was extended to other ornamental bits such as harness fittings and weapon fittings. After the first century A.D., the copper, tin, and zinc became intermingled; little care was taken to keep the alloys pure.

Other studies of Gallo-Roman copper-base metal use demonstrate that the Gauls, especially in central France and the Rhineland, preferred to use a great deal of zinc in religious statues.

The Romans preferred to use the various alloys in ways different from the Gauls. After 23 B.C., brass was reserved for coinage and military fittings, all functions ostentatiously associated with imperial power. Little zinc was used in Roman votive statues. After the first century A.D., the amount of zinc in the coins declined, and again, as with the Gauls, there seems to have been a casual mixing of remelted alloys. The different valuations of the alloys were largely abandoned, and brass seems to have become merely one more useful industrial metal.

For both the Gauls and the Romans, brass, or orichalcum, was rooted in legend. Chapter 3 describes the high valuation put on this rare, gold-mimicking metal before 100 B.C. In the early Imperium some of this prestige was probably retained, though the cementation process could, with some effort, produce a great deal of brass. Much of this high valuation was probably due not only to the historical value of brass, but also to the psychological effect of Augustus's reform of the currency in 23 B.C. The bronze coinage of the Republic had been debased and was essentially worthless; Augustus's new coinage was solid, reliable—and made of brass. The use of brass for Roman military fittings may have been an attempt to associate the army with this sound and trusted aspect of the governmental system.

We may conclude that both the Gauls and the Romans set a fairly high value on brass in the first centuries B.C. and A.D., and they both reserved it for certain value-laden purposes. In Rome, as described above, it was coinage and the military; in Gaul it was largely fibulae.

Gallic society was a prestige-goods society (Cunliffe 1988:198), one where status was displayed and negotiated through the competitive possession and bestowal of precious items. After torcs, the flashiest and most visible ornaments were the fibulae, worn as they were by both sexes in prominent places on the chest and shoulders. We do not know the various meanings fibulae must have had in Gallic society, but it is clear from their ubiquity and rapid changes in form that they must have been a prime focus for display, and it is for them that this gold-seeming metal, which acquired status as well from its history, was reserved.

Another clue to the high valuation of brass among the Gauls is found in the figurine analyses. The statuettes from central France and the Rhineland have noticeably more zinc than figurines made elsewhere in Gaul, already a high-zinc use area. Since most of these statues are thought to be religious or votive in nature, we may hypothesize that the Gauls connected zinc and brass with religious artifacts.

But values, as seen in Chapter 2, do not reside solely in the finished object. Values are also expressed in the manufacture of the artifact, and brass manufacturing practices in Gaul of the first centuries B.C. and A.D. clearly show how the various alloys were separated. The debris pieces at the Titelberg were all composed of tin bronze; there is no evidence that either copper or brass was being worked in that building at any time. In Alesia, two first century A.D. workshops were found that specialized solely in brassworking; bronze- and copper-working areas were located elsewhere in the city.

Specialization is also shown in the source of the copper. The compositional analyses of the trace elements in the Titelberg samples demonstrate that in the period ca. 50 B.C.–A.D. 70, the copper used to make brass came from a different ore source than the rather heterogeneous copper ores that supplied the copper for the bronze and impure copper artifacts. Carter and Buttrey (1977) have similar findings for the copper and brass coins of Augustus and Tiberius; the copper used in the brass coins had distinctly lower nickel percentages than the metal used in the copper coins. This suggests that this specialization of source was a phenomenon widespread in the western Empire, at least in the first 150 years after the conquest of Gaul.

To put all this together, we recall Pfaffenberger's concept of "sociotechnical systems," systems comprised of techniques and material culture, the social coordination of labor, and the social meaning of technological activities (Pfaffenberger 1992). It is suggested that in both Gaul and the Roman Empire as a whole, a new sociotechnical system developed, one built on the manufacturing techniques, labor organization, and social meanings of brass. The manufacture of brass required the exploitation of a new kind of ore, that of zinc, and the use of the specialized technique of cementation. This manufacture took place in specialized workshops, and when the metal was produced, it was crafted into artifacts in specialized smithies, at least in Gaul, and reserved for specialized functions. In Gaul, brass largely replaced bronze and copper in fibulae and ornaments; in the Mediterranean brass replaced bronze and copper for coinage and military fittings. In both cases, while the specific values and information that brass contained differed, the metal conveyed much meaning. Though there is no direct evidence to this effect, we can infer that specialized brass smiths, brass smelters, and zinc miners came into being. The rules and regulations faced by, or developed by, these miners and smiths must have differed for bronze and copper smiths as well. Though we have little information on this point, the Imperial government had a considerable interest in zinc mines because of the use of zinc in coinage, and these mines may well have been worked or monitored by soldiers.

The technological system surrounding brass encompassed far more than the mere appearance of a new and striking metal; it was accompanied by changes in technology, mining, labor coordination, and values. This is a classic case of the development of a new sociotechnical system, one, furthermore, that did not, as far as we can tell, replace the existing system built up around the manufacture and use of bronze and copper. Brass replaced bronze and copper in a number of uses in both Gaul and Italy, but the latter two continued to be made and used, and, judging from the continued use of bronze in pins and copper in rivets, they conveyed their own meanings.

These distinct new sociotechnical systems seem to have largely vanished in the second century A.D. In Gaul, fibulae are less frequently found in the archaeological record. The use of zinc spreads to many other artifact categories; ternary and quaternary alloys are common. There is much less brass being manufactured, and the average zinc levels in coins and artifacts decline. Military fittings, as far as we know, are no longer of brass; analyses of post–first century fittings are nonexistent. Since we have no workshop excavations dating to the second century and later, we cannot tell if there is still specialization of smithies, but since so much brass and bronze are being melted together, it seems unlikely. In Gaul and Italy, the special values attached to brass seem to fade, until there is little difference between the two areas. They seem to merge into one common sociotechnical system, the result of mutual homogenization.

Romanization and metalworking

By the time of the Caesarian invasion, Belgic Gaul, like the rest of Gaul, had had centuries of contact with the Mediterranean world. Some Belgic men had traveled as mercenaries, others as traders; small quantities of Greek, Etruscan, or Roman luxury goods had appeared in Gallic graves since 600 B.C. It is notable that, in *De Bello Gallico*, Caesar never mentions any difficulty in communicating with the Gauls, enemy or ally. Since he certainly spoke no Gallic language, presumably the ability to speak Latin or Greek was widespread enough in Gaul for him to have no difficulty finding translators. The only mention of translators is in Book 1.47, where he notes that he sent a Romanized Gaul to speak in Gallic to Ariovistus, a German leader.

In addition, there was no large technological gap between the Gauls and the Romans. Where the Romans were particularly skilled at stone working and the building of roads and aqueducts, the Gauls were ahead of the Romans in many aspects of ironworking and the technology of farm tools and wagons. It is true, of course, that the Romans conquered Gaul, but their advantage lay in their organization and their unity. The Roman army was disciplined, with stamina for the long battle and trained to fight together; the Gauls were terrifying in the first charge but faltered in the long run. The Romans, despite their internal struggles, presented a united front to the enemy; the Gallic tribes would not unite, even in the face of a common danger. Nothing helped Caesar more in his conquest than the intertribal dissensions of the Gauls.

After the Conquest, the legions remained in Gaul, to monitor the Gauls but mostly to protect Rome's new property from the Germans across the Rhine. The ineffective attempt of Augustus to conquer Free Germany, which ended in A.D. 16, led to the permanent positioning of troops along the Rhine River, which became the frontier.

Agents of cultural transmission in Gaul

Acculturation theory (Bee 1974; SSRC 1954) describes several factors that affect what kind of information is being transmitted to the receptor culture. One of the most important is, with whom among the donor culture are the receivers in contact? This has a profound impact on the kinds of information that are transmitted. In the case of Belgic Gaul, initially there would have been a few Roman traders, but the bulk of the Mediterranean goods and information would probably have been delivered through Gallic *civitates* closer to the Mediterranean. Returning mercenaries would also have brought information acquired in a mercenary context, which means that it was rather limited. Mercenaries served in units formed of their own countrymen, most of whom probably spoke the language of wherever they were serving imperfectly if at all. Most of the contact the mercenaries would have had with Mediterranean peoples would come either from guarding them or killing them. Other contacts would have come from the men who hired the mercenaries, the merchants who sold them food and wine, and the women with whom they slept. No matter how curious and observant a young Celtic mercenary would have been, the amount of accurate (or at least emic) information about Roman, Greek, and Asiatic culture that he could have picked up was decidedly limited. This limited and inevitably garbled information would be taken back to Gaul and there filtered through the culture and assumptions of the untraveled segment of society.

Though there is little sign of the seven years of battle and invasion in the archaeological record, it is probable that Caesar's invasion caused serious disruption in Gallic society. Huge numbers of Gauls were killed and enslaved. (Plutarch estimates that one million were killed and another million enslaved [Drinkwater 1983:119]; these figures are suspect, but the total must still have been very large.) The civil wars in Italy that followed the return of Caesar led to Gallic *civitates* being bled of their young men to form auxiliary troops: many tribes were required to supply large numbers of soldiers. The Conquest, then, not only wiped out the most anti-Roman of the Gauls but rather forcibly acculturated large groups of young men (ibid.).

After the Conquest, the occupying army and the garrison on the Rhine frontier would have been the main sources of information about Roman culture, especially for tribes such as the Treveri, who were situated near the frontier. The army did not merely occupy, it also needed to be supplied, and the troops spent their pay in Gaul. Therefore the main conduits of information, especially in the first century after the Conquest, would have been lower- or middle-class males, from both urban and rural origins, probably badly educated and not very familiar with the upper reaches and scholarship of their own culture. The groups of Gauls with which these men would have had the most interaction would have been the farmers, merchants, and craftworkers who would have supplied their material needs, the women whom they would inevitably have had as concubines, and the children these women would have had by the soldiers.

Officers and governmental officials would have dealt more with the nobles, who were probably already fairly pro-Roman, since they had survived. Many of these officials would probably have brought their families and servants along; one could expect that noble Roman ladies would have had an influence on Gallic women far out of proportion to their numbers. They would have been potent sources of information on dress, household goods, cookery, and domestic architecture, as well as religious practices and the conduct of female-male relationships. Despite this, the vast majority of information was still transmitted by military men: occupiers, customers, and protectors from the Germans.

At the time of the Conquest, the Gauls were in a stage where their sociopolitical complexity was rapidly growing. Proto-urban sites stretched from Britain to Moravia. *Civitates* in central France and Switzerland were abolishing kingships and setting up elected aristocratic oligarchies (the same process that had taken place in many Greek cities and then Rome itself some centuries earlier). From the omnipresent fortified sites, the weapons in graves, and the widespread La Tène material culture, one can assume that this was a time of continual intertribal warfare and competitive emulation. This scene fits very well van der Leeuw's (1983:18) description of an "expanding system," one where new levels of abstraction and information have just been added, where change is rapid, and where the system is easily capable of adding new information and new structures. This may explain the rapidity with which Gaul adapted to the Roman administration, and the degree to which it adopted the Roman language and material culture. This is of course impossible to test, but it is possible that if Rome had invaded two centuries earlier, before the visible signs of this expanding system appear in the record, the extent of ultimate Romanization would have been much less.

To sum up, after centuries of rather casual contact that only served to transmit material goods and limited cultural information, Gaul was brought forcibly into the administrative and military structure of the Roman Empire at a particular stage in its development where it was most able to absorb new material. In the absence of deliberate Imperial mechanisms for Romanizing the barbarians, we must reconstruct by whom most of the cultural information was conveyed and what they were conveying.

Gaul and its cultural interactions

It is usually assumed in contact studies that acculturation takes place when two cultures interact over the long term, and that almost invariably these cultures are different in power and technology. Both the history of Gaul and the results produced by this study suggest that this model is oversimplified. A diagram of the various cultural interactions engaged in by the Treveri in particular is presented in Fig. 49. It is a complex diagram, like its subject, and needs further discussion. At the heart is a circle that encompasses the *civitates* of the Treveri. Inside the Treveri are factions, as in all Gallic tribes; there are pro-Roman and anti-Roman factions, there is conflict between client and patron, and there are power manipulations unrelated to the Romans.

The Treveri are also in peer-polity interactions with neighboring tribes. Peer-polity interactions take place between polities that are roughly equal in power and share some common cultural traditions (Renfrew 1986). The "competitive emulation" between the polities, which was also a notable feature of Gallic life, drives the process both of increasing sociopolitical complexity and the rapid spread of material culture and ideas (Renfrew 1986). Though this aspect of post-Conquest Gallic life is usually ignored, one can hypothesize that even after the Conquest the Treveri were intensely interested in and influenced by what was happening in other *civitates*. This was acknowledged by Rome when, at the early stages, it sought to govern by enrolling as allies two or three of the most influential tribes (Drinkwater 1983:19). One can suggest that those practices and material goods adopted by the Aedui or the Remi were much more likely to be adopted by the Treverans, and vice versa; the Treverans, after all, were a powerful and wealthy tribe. Spokes that connect the Treveri to these other tribes are shown on the diagram, as well as the circle that delimits (in an oversimplified way) the Gallic cultural sphere.

The Germans occupy an ambiguous position. They are mostly outside the Three Gauls, but they have genealogical connections with the Treveri, and the incursions of Ariovistus and his Suebi tribe in the Gallic Wars are merely one example of the constant pressure of the Germans on the border. The German influence is most pronounced near the frontier, including the territory of the Treveri, but they are both inside and outside the area of intra-Gallic interactions. A dotted line on Fig. 49 shows this ambiguity.

All these tribes interact after the Conquest with the Roman Empire, in the form of soldiers, administrators, administrators' families, merchants, and tax collectors. Rome deals with Gaul as a whole when it formulates policies, and with each tribe individually in the field. The interactions with individual *civitates* depend a great deal on each tribe's nearness to the frontier, history of rebellion or alliance, and natural resources, such as mines.

Naturally all the elements in the Roman circle are not always unified, either; the policies of the army commanders may be at odds with the self-interest of the merchants and contractors or procurators. Since all of these were operating in Gaul at the same time, a certain confusion of results and information transmitted should be expected.

The largest circle delimits the northern Mediterranean, Gaul, Britain, and Germany as one large European culture system. This system is formed by centuries of interaction. It interacts as a system with other areas, such as Asia, the steppes, and North Africa. This does not mean that Europe as a whole, however that is defined, necessarily interacts with these different areas, but that any part of Europe can interact with, and acquire information, behaviors, and artifacts from, these areas. Anything acquired from outside the European culture area tends to spread fairly rapidly throughout Europe, under the constraints of each individual culture or contact situation, to all parts of the area.

We see the workings of this complex set of interactions when we consider the new sociotechnical system of brass manufacture and use. The findings of this study suggest that the knowledge and high valuation of brass did not spread from Rome to Belgic Gaul. The Treveran area worked with brass earlier than the Romans did, almost as early as the inventors of the cementation process in Asia Minor. Contrarily, there is certainly no evidence to suggest that the brass spread from Gaul to Rome. What seems to have happened is that Belgic Gaul and the Roman Republic/Empire acquired the knowledge of cementation brass independently, from Asia Minor. As discussed above, Belgic Gaul and Rome both built new sociotechnical systems around brass (though the evidence for that is clearer in Gaul). The metal was reserved for special purposes, acquired from special ores, and worked in special workshops. Bronze and copper, too, had special purposes and specialized workshops, as on the Titelberg.

So the Treverans and the Romans both had special new sociopolitical systems to deal with this brass. But at

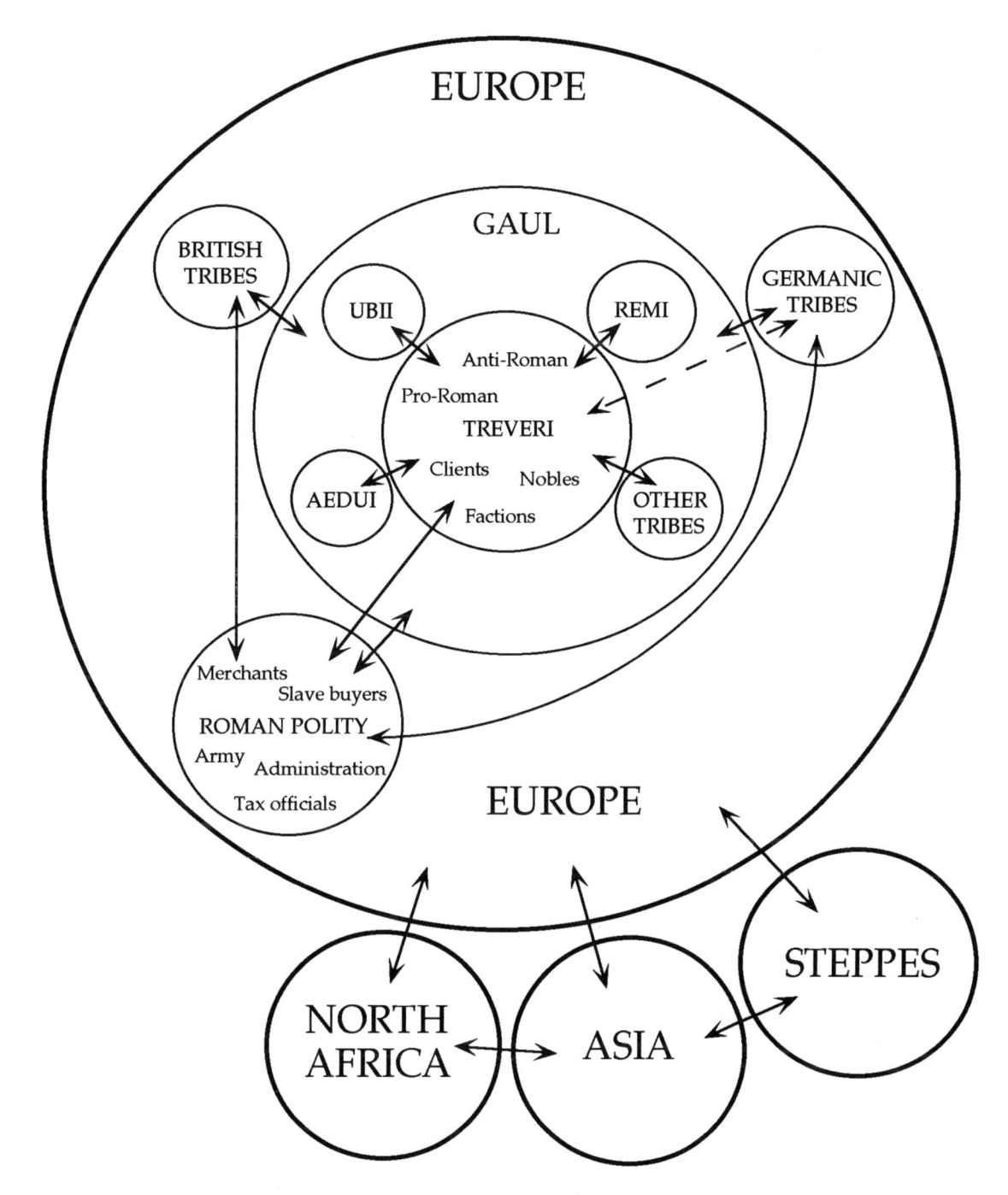

Fig. 49:
The Treveri and their various spheres of interaction.

the end of the first century B.C. these systems seem to have dissolved. Fresh brass was seldom made, bronze and brass and copper and lead were intermixed in many artifact classes. Even if, as Craddock (1978) maintains, this mixing is for the purpose of making an alloy of general utility, it still indicates that the meanings attached to bronze and brass have changed. This opportunistic mixing also suggests that the workshops that specialized in brass or bronze have merged. Characteristic regional recipes are still seen, but that mostly seems to be because the smiths are remelting scrap produced by the old recipes.

What is striking is that Belgic Gaul and Rome are following the same trend. We do not know if the trend came from Rome and was followed by increasingly Romanized craftworkers, or if it was a case of convergent evolution, as it were. The result was a homogenization of copper-base metal use; both the distinctive sociotechnical systems disappeared.

This result suggests that any discussion of Romanization or indeed any contact situation needs to be widened to include internal interactions, and especially the interactions taking place by both cultures with societies outside their relationship. The changes in metalworking found at the Titelberg cannot be explained either by theories of retention of native metallurgical habits or by diffusion from the Roman Empire. Instead, it was a result of diffusion of artifacts, knowledge, and values acquired from Asia Minor, which were built into different sociotechnical systems in Gaul and Rome. The end result was a homogenization of copperworking technology and values over Roman Europe, with the retention of some regional recipes and, perhaps, some values. In this respect Gaul and Rome joined together to participate in something new.

Notes

11. In 45 B.C., C. Clovius issued brass coins for Julius Caesar. This issue is not well known; numismatists have paid them little attention.

REFERENCES CITED

Agache, R. 1978. *La Somme pré-romaine et romaine*. Mémoires de la Société des Antiquaires de Picardie 24. Amiens.

Alexander, J. 1981. The Coming of Iron-Using to Britain. In *Frühes Eisen in Europa*, ed. H. Haefner, pp. 57–67. Verlag Peter Meili, Schaffhausen.

Allen, D. F. 1980. *The Coins of the Ancient Celts*, ed. D. Nash. Edinburgh University Press, Edinburgh.

Appadurai, A. (ed.). 1986. *The Social Life of Things*. Cambridge University Press, Cambridge.

Bachmann, H.-G. 1982. *The Identification of Slags from Archaeological Sites*. Institute of Archaeology, University of London, Occasional Paper 6.

Barley, N. 1986. *Ceremony: An Anthropologist's Misadventures in the African Bush*. Henry Holt, New York.

Basalla, G. 1988. *The Evolution of Technology*. Cambridge University Press, Cambridge.

Bayley, J. 1984. Copper Alloys and Their Use in Iron Age and Roman Britain. Paper presented at 1984 Symposium on Archaeometry, Smithsonian Institution, Washington, DC, 14–18 May 1984.

⸺ 1985. Brass and Brooches in Roman Britain. *MASCA Journal* 3(6):189–191.

⸺ 1990. The Production of Brass in Antiquity with Particular Reference to Roman Britain. In *2000 Years of Zinc and Brass*, ed. P. T. Craddock, pp. 7–28. British Museum Occasional Papers 50. London.

Bayley, J., and S. A. Butcher. 1981. Variations in Alloy Composition of Roman Brooches. *Revue d'archéometrie* 1 (Supplement):7–28.

Beck, F., M. Menu, T. Berthoud, and L.-P. Hurtel. 1985. Métallurgie des bronzes. In *Recherches gallo-romaines* I, pp. 69–140. Laboratoire des Recherches des Musées de France, Paris.

Bee, R. L. 1974. *Patterns and Processes: An Introduction to Anthropological Strategies for the Study of Sociocultural Change*. The Free Press, New York.

Benedict, R. 1948. Anthropology and the Humanities. *American Anthropologist* 30:585–593.

Bijker, W. E. 1987. The Social Construction of Bakelite: Towards a Theory of Invention. In *The Social Construction of Technological Systems*, ed. W. E. Bijker, T. P. Hughes, and T. Pinch, pp. 159–187. MIT Press, Cambridge, MA.

Binford, L. 1962. Archaeology as Anthropology. *American Antiquity* 28(2):217–225.

⸺ 1965. Archaeological Systematics and the Study of Culture Process. *American Antiquity* 31:203–210.

Bintliff, J. 1984. Iron Age Europe in the Context of Social Evolution from the Bronze Age Through to Historic Times. In *European Social Evolution*, ed. J. Bintliff, pp. 157–226. University of Bradford, Bradford.

Blair, C. E. 1992. The Scale and Organization of the Iron Industry at Celtic Oppida: As Characterized by the Oppidum at Kelheim, Bavaria. Ph.D. diss., University of Minnesota.

Bleed, P. 1993. Product History as Archaeology. *Archeomaterials* 7(1):177–180.

Boas, F. 1940. *Race, Language, and Culture*. Macmillan, New York.

Bordes, F. 1972. *A Tale of Two Caves*. Harper and Row, New York.

Bordes, F., and D. de Sonneville-Bordes. 1970. The Significance of Variability in Paleolithic Assemblages. *World Archaeology* 1:61–73.

Brumfiel, E. M. 1987. Elite and Utilitarian Crafts in the Aztec State. In *Specialization, Exchange and Complex Societies*, ed. E. M. Brumfiel and T. K. Earle, pp. 102–118. Cambridge University Press, Cambridge.

Brumfiel, E. M., and T. K. Earle. 1987. Specialization, Exchange, and Complex Societies: An Introduction. In *Specialization, Exchange and Complex Societies*, ed. E. M. Brumfiel and T. K. Earle, pp. 1–9. Cambridge University Press, Cambridge.

Buchwald, V. F., and P. Leisner. 1990. A Metallurgical Study of 12 Prehistoric Bronze Objects from Denmark. *Journal of Danish Archaeology* 9:64–102.

Butler, J. J., and J. D. van der Waals. 1966. Bell Beakers and Early Metal-Working in the Netherlands. *Palaeohistoria* 12:42–139.

Butts, A. 1954. *Copper*. Reinhold, New York.

Caesar, G. J. 1980. *The Battle for Gaul*, transl. A. and P. Wiseman. Book Club Associates, London.

Caley, E. R. 1964. *Orichalcum and Related Ancient Alloys*. American Numismatic Society 151, New York.

⸺ 1970. Chemical Composition of Greek and Roman Statuary Bronzes. In *Art and Technology, A Symposium on Classical Bronzes*, ed. S. Doeringer, D. G. Mitten, and A. Steinberg, pp. 37–49. MIT Press, Cambridge, MA.

Callon, M. 1987. Society in the Making: The Study of Technology as a Tool for Sociological Analysis. In *The Social Construction of Technological Systems: New Directions in the Sociology and History of Technology*, ed. W. B. Bijker, T. P. Hughes, and T. Pinch, pp. 83–103. MIT Press, Cambridge, MA.

Carter, G. F. 1971. Compositions of Some Copper-Based Coins of Augustus and Tiberius. In *Science and Archaeology*, ed. R. H. Brill, pp. 114–129. MIT Press, Cambridge, MA.

───── 1978. Chemical Compositions of Copper-Based Roman Coins. Augustan Quadrantes, ca. 9–4 B.C. In *Archaeological Chemistry*, Vol. 2. Advances in Chemistry Series 171, pp. 347–377. American Chemical Society, Washington, DC.

Carter, G. F., and T. V. Buttrey. 1977. Chemical Compositions of Copper-Based Roman Coins, II: Augustus and Tiberius. *American Numismatist Society: Museum Notes* 22:49–66.

Chase, W. T. 1974. Comparative Analysis of Archaeological Bronzes. In *Archaeological Chemistry*, ed. C. W. Beck, pp. 148–185. American Chemical Society, Washington, DC.

Childe, V. G. 1929. *The Danube in Prehistory*. Oxford University Press, Oxford.

───── 1944. Archaeological Ages as Technological Stages. *Journal of the Royal Anthropological Institute* 74:7–24.

───── 1958. Retrospect. *Antiquity* 22:69–74.

Childs, S. T. 1991a. Iron as Utility or Expression: Reforging Function in Africa. In *Metals in Society: Theory Beyond Analysis*, ed. R. M. Ehrenreich, pp. 57–68. MASCA Research Papers in Science and Archaeology, Vol. 8(2). MASCA, University of Pennsylvania Museum, Philadelphia.

───── 1991b. Transformations: Iron and Copper Production in Central Africa. In *Recent Trends in Archaeometallurgical Research*, ed. P. D. Glumac, pp. 33–46. MASCA Research Papers in Science and Archaeology, Vol. 8(1). MASCA, University of Pennsylvania Museum, Philadelphia.

───── 1994. Society, Culture, and Technology in Africa: An Introduction. In *Society, Culture, and Technology in Africa*, ed. S. T. Childs, pp. 6–14. MASCA Research Papers in Science and Archaeology 11 (Supplement). MASCA, University of Pennsylvania Museum, Philadelphia.

Childs, S. T., and D. Killick. 1993. Indigenous African Metallurgy: Nature and Culture. *Annual Review of Anthropology* 22:317–337.

Clark, J. G. D. 1952. *Prehistoric Europe: The Economic Base*. Methuen, London.

Cleere, H. 1993. Archaeometallurgy Comes of Age. *Antiquity* 67:175–178.

Coghlan, H. H. 1951. *Notes on the Prehistoric Metallurgy of Copper and Bronze in the Old World*. Pitt-Rivers Museum Occasional Papers on Technology 4. Oxford University Press, Oxford.

Coles, J. M., and A. F. Harding. 1979. *The Bronze Age in Europe*. Methuen, London.

Collis, J. 1984a. *The European Iron Age*. Schocken Books, New York.

───── 1984b. *Oppida*. University of Sheffield, Sheffield.

Condamin, J., and S. Boucher. 1973. Recherches techniques sur des bronzes de Gaule romaine IV. *Gallia* 31:157–178.

Costin, C. L. 1991. Craft Specialization: Issues in Defining, Documenting, and Explaining the Organization of Production. In *Archaeological Method and Theory*, Vol. 3, ed. Michael B. Schiffer, pp. 1–55. University of Arizona Press, Tucson.

Cowan, R. S. 1983. *More Work for Mother: The Ironies of Household Technology from the Open Hearth to the Microwave*. Basic Books, New York.

Craddock, P. T. 1978. The Composition of the Copper Alloys Used by the Greek, Etruscan, and Roman Civilizations, Part III. *Journal of Archaeological Science* 5(1):1–16.

───── 1985. Three Thousand Years of Copper Alloys: From the Bronze Age to the Industrial Revolution. In *Applications of Science in Examination of Works of Art*, ed. P. A. England and L. van Zelst, pp. 59–68. Boston Museum of Fine Arts, Boston.

───── 1989. The Scientific Investigation of Early Mining and Metallurgy. In *Scientific Analysis in Archaeology*, ed. J. Henderson, pp. 178–212. UCLA Institute of Archaeology, Los Angeles.

Craddock, P. T. (ed.). 1990. *2000 Years of Zinc and Brass*. British Museum Occasional Papers 50. London.

Craddock, P. T., A. M. Burnett, and K. Preston. 1980. Hellenistic Copper Base Coinage and the Origins of Brass. In *Scientific Studies in Numismatics*, ed. W. A. Oddy, pp. 53–64. British Museum Occasional Papers 18. London.

Craddock, P. T., I. C. Freestone, N. H. Gale, N. D. Meeks, B. Rothenberg, and M. S. Tite. 1985. The Investigation of a Small Heap of Silver Smelting Debris from Rio Tinto, Huelva, Spain. In *Furnaces and Smelting Technology in Antiquity*, ed. P. T. Craddock and M. J. Hughes, pp. 199–217. British Museum Occasional Papers 48. London.

Craddock, P. T., I. C. Freestone, L. K. Gujar, A. P. Middleton, and L. Willies. 1990. Zinc in India. In *2000 Years of Zinc and Brass*, ed. P. T. Craddock, pp. 29–72. British Museum Occasional Papers 50. London.

Craddock, P. T., and A. R. Giumlia-Mair. 1988. Problems and Possibilities for Provenancing Bronzes by Chemical Composition, with Special Reference to Western Asia and the Mediterranean in the EIA. In *Bronzeworking Centres of Western Asia, c. 1000–539 B.C.*, ed. J. Curtis, pp. 317–328. Kegan Paul, London.

Craddock, P., and J. Lambert. 1985. The Composition of the Trappings. *Britannia* 16:162–164.

Crawford, M. H. 1985. *Coinage and Money Under the Roman Republic*. University of California Press, Berkeley.

Crumley, C. L. 1974. *Celtic Social Structure: The Generation of Archaeologically Testable Hypotheses from Literary Evidence*. University of Michigan Press, Ann Arbor.

——— 1987. Celtic Settlement Before the Conquest: The Dialetics of Landscape and Power. In *Regional Dynamics: Burgundian Landscapes in Historical Perspective*, ed. C. L. Crumley and W. H. Marquardt, pp. 403–429. Academic Press, San Diego.

Cunliffe, B. 1988. *Greeks, Romans and Barbarians: Spheres of Interaction*. Methuen, New York.

De Barros, P. 1988. Societal Repercussions of the Rise of Large-Scale Traditional Iron Production: A West African Example. *The African Archaeological Review* 6:91–113.

Deetz, J. F. 1965. *The Dynamics of Stylistic Change in Arikara Ceramics*. University of Illinois Series in Anthropology 4. Urbana.

Demoule, J.-P., and M. Ilett. 1985. First-Millennium Settlement and Society in Northern France: A Case Study from the Aisne Valley. In *Settlement and Society: Aspects of Western European Prehistory in the First Millennium B.C.*, ed. T. C. Champion and J. V. S. Megaw, pp. 193–221. St. Martin's Press, New York.

Dietler, M. 1989. Greeks, Etruscans, and Thirsty Barbarians: Early Iron Age Interaction in the Rhône Basin of France. In *Centre and Periphery: Comparative Studies in Archaeology*, ed. T. C. Champion, pp. 127–141. Unwin Hyman, London.

——— 1990. Driven by Drink: The Role of Drinking in the Political Economy and the Case of Early Iron Age France. *Journal of Anthropological Archaeology* 9:352–406.

Dobres, M.-A., and C. R. Hoffman. 1994. Social Agency and the Dynamics of Prehistoric Technology. *Journal of Archaeological Method and Theory* 1(3):211–258.

Dowdle, J. E. 1987. Road Networks and Exchange Systems in the Aeduan *Civitas*, 300 B.C.–A.D. 300. In *Regional Dynamics: Burgundian Landscapes in Historical Perspective*, ed. C. L. Crumley and W. H. Marquandt, pp. 265–294. Academic Press, San Diego.

Drinkwater, J. F. 1983. *Roman Gaul: The Three Provinces 58 BC–AD 260*. Croom Helm, London.

Dunnell, R. C. 1993. Why Archaeologists Don't Care About Archaeometry. *Archeomaterials* 7:161–165.

Dunning. F. W., and A. M. Evans (eds.). 1986. *Mineral Deposits of Europe*. Vol. 3: *Central Europe*. Institution of Mining and Metallurgy and the Mineralogical Society, London.

Earle, T. K. 1981. Comment on Rice. *Current Anthropology* 22(3):230–231.

Ehrenreich, R. M. 1985. *Trade, Technology and the Ironworking Community of Southern Britain in the Iron Age*. BAR British Series 144. British Archaeological Reports, Oxford.

——— 1991. Metalworking in Iron Age Britain: Hierarchy or Heterarchy? In *Metals in Society: Theory Beyond Analysis*, ed. R. M. Ehrenreich, pp. 69–80. MASCA Research Papers in Science and Archaeology 8(2). MASCA, University of Pennsylvania Museum, Philadelphia.

Eisenstadt, S. N., and L. Roniger. 1980. Patron-Client Relations as a Model of Structuring Social Exchange. *Comparative Studies in Society and History* 22(1):42–77.

Epstein, S. M. 1993. Cultural Choice and Technological Consequences: Constraint of Innovation in the Late Prehistoric Copper Smelting Industry of Cerro Huaringa, Peru. Ph.D diss., University of Pennsylvania.

Fitzpatrick, A. 1985. The Distribution of Dressel 1 Amphorae in North-West Europe. *Oxford Journal of Archaeology* 4(3):305–340.

Fleming, S. J. 1985. The Application of PIXE Spectrometry to Bronze Analysis: Practical Considerations. *MASCA Journal* 3(5):142–149.

Fleming, S. J., and C. P. Swann. 1993. Recent Applications of PIXE Spectrometry in Archaeology. Part I: Observations on the Early Development of Copper Metallurgy in the Old World. *Nuclear Instruments and Methods in Physics Research* B75:440–444.

Forbes, R. J. 1950. *Metallurgy in Antiquity*. E. J. Brill, Leiden.

——— 1955–64. *Studies in Ancient Technology*. E. J. Brill, Leiden.

Franklin, U. M. 1983. The Beginnings of Metallurgy in China: A Comparative Approach. In *The Great Bronze Age of China: A Symposium*, ed. G. Kuwayama, pp. 94–98. Los Angeles County Museum of Art, Los Angeles.

Gale, N., and Z. Stos-Gale. 1982. Bronze Age Copper Sources in the Mediterranean: A New Approach. *Science* 216:11–19.

Geselowitz, M.N. 1988. The Role of Iron Production in

the Formation of an "Iron Age Economy" in Central Europe. *Research in Economic Anthropology* 10:225–255.

Gille, B. 1978. *Histoire des techniques*. Gallimard, Paris.

Glumac, P. D. (ed.). 1991. *Recent Trends in Archaeometallurgical Research*. MASCA Research Papers in Science and Archaeology 8(1). MASCA, University of Pennsylvania Museum, Philadelphia.

Glumac, P. D., and J. A. Todd. 1991. Early Metallurgy in Southeast Europe: The Evidence for Production. In *Recent Trends in Archaeometallurgical Research*, ed. P. D. Glumac, pp. 8–20. MASCA Research Papers in Science and Archaeology 8(1). MASCA, University of Pennsylvania Museum, Philadephia.

Grant, M. 1946. *From Imperium to Auctoritas*. Cambridge University Press, Cambridge.

Griffin, J. B. 1946. Culture Change and Continuity in Eastern United States. In *Man in Northeastern North America*, ed. F. Johnson, pp. 37–95. Peabody Foundation, Andover, MA.

Griffith, G. T. 1935. *The Mercenaries of the Hellenistic World*. Cambridge University Press, Cambridge.

Gruel, K., F. Widemann, M. Barral, W. Fedoroff, J. Leres, C. Neskovic, M. Piolet, G. Revel. 1979. Typological and Analytical Study of Celtic Coins from the Trebry Hoard. Proceedings of the 18th International Symposium on Archaeometry and Archaeological Prospection. *Archaeo-Physika* 10:50–67.

Haffner, A. 1989. Das Gräberfeld von Wederath-Belginium vom 4. Jahrhundert vor bis zum 4. Jahrhundert nach Christi Geburt. In *Gräber-Spiegel des Lebens: Zum Totenbrauchtum der Kelten und Römer*, ed. A. Haffner, pp. 37–129. Rheinisches Landesmuseum Trier, Mainz am Rhein.

Halleux, R. 1973. L'orichalque et le laiton. *L'antiquité classique* 42:64–81.

Hamilton, E. 1991. Metallurgical Analysis and the Bronze Age of Bohemia: Or, Are Cultural Alloys Real? *Archaeomaterials* 5:75–89.

_____ 1995. Was There Ever a Roman Conquest? In *Different Iron Ages: Studies on the Iron Age in Temperate Europe*, ed. J. D. Hill and C. Cumberpatch, pp. 37–44. BAR International Series 602. British Archaeological Reports, Oxford.

Hamilton, E., C. P. Swann, and S. J. Fleming. 1994. Roman Influences on Metalworking at the Titelberg (Luxembourg): Compositional Studies Using PIXE Spectrometry. *Nuclear Instruments and Methods in Physics Research* B85:856–860.

Harris, M. 1968. *The Rise of Anthropological Theory*. Thomas Y. Crowell, New York.

_____ 1979. *Cultural Materialism: The Struggle for a Science of Culture*. Vintage Books, New York.

Hartmann, A., and E. Sangmeister. 1972. The Study of Prehistoric Metallurgy. *Angewandte Chemie* 11(7):620–629.

Haselgrove, C. 1990. The Romanization of Belgic Gaul: Some Archaeological Perspectives. In *The Early Roman Empire in the West*, ed. T. Blagg and M. Millett, pp. 45–71. Oxbow Books, Oxford.

Hatt, J.-J. 1970. *Celts and Gallo-Romans*. Nagel Publishers, Geneva.

Haudricourt, A.-G. 1962. Domestication des animaux, culture des plantes et traitement a'autrui. *L'homme* 2:40–50.

_____ 1988. *La technologie, science humaine: Recherche d'histoire et d'ethnologie des techniques*. Editions de la Maison des Sciences de l'Homme, Paris.

Hegmon, M. 1992. Archaeological Research on Style. *Annual Review of Anthropology* 21:517–536.

Heskel, D., and C. C. Lamberg-Karlovsky. 1980. An Alternative Sequence for the Development of Metallurgy: Tepe Yahya, Iran. In *The Coming of the Age of Iron*, ed. T. A. Wertime and J.D. Muhly, pp. 229–265. Yale University Press, New Haven.

Higgins, R. 1980. *Greek and Roman Jewellery*. Methuen, London.

Higgs, E. S., and C. Vita-Finzi. 1972. Prehistoric Economies: A Territorial Approach. In *Papers in Economic Prehistory*, ed. E. S. Higgs, pp. 27–36. Cambridge University Press, Cambridge.

Hodder, I. 1982. *Symbols in Action: Ethnoarchaeological Studies of Material Culture*. Cambridge University Press, Cambridge.

_____ 1984. Burials, Houses, Women and Men in the European Neolithic. In *Ideology, Power and Prehistory*, ed. D. Miller and C. Tilley, pp. 51–68. Cambridge University Press, Cambridge.

_____ (ed.). 1989. *The Meaning of Things: Material Culture and Symbolic Expression*. Unwin Hyman, Boston.

Hodges, H. 1976. *Artifacts*. Humanities Press, NJ.

Hosler, D. 1988. Ancient West Mexican Metallurgy: South and Central American Origins and West Mexican Transformations. *American Anthropologist* 90(4):832–855.

Hounshell, D. 1984. *From the American System to Mass Production, 1800–1932*. Johns Hopkins University Press, Baltimore.

Hudson, M. 1973. *Structure and Metals*. Hutchinson Educational, London.

Hughes, T. P. 1983. *Networks of Power: Electrification in Western Society, 1880–1930*. Johns Hopkins University Press, Baltimore.

_____ 1987. The Evolution of Large Technological

Systems. In *The Social Construction of Technological Systems*, ed. W. E. Bijker, T. P. Hughes, and T. Pinch, pp. 51–82. MIT Press, Cambridge, MA.

Jackson, J. S. 1980. Bronze Age Copper Mining in Counties Cork and Kerry, Ireland. In *Scientific Studies in Early Mining and Extractive Metallurgy*, ed. P. T. Craddock, pp. 9–30. British Museum Occasional Papers 20. London.

Jovanović, B. 1980. Primary Copper Mining and the Production of Copper. In *Scientific Studies in Early Mining and Extractive Metallurgy*, ed. P. T. Craddock, pp. 31–40. British Museum Occasional Papers 20. London.

Junghans, S., E. Sangmeister, and M. Schroeder. 1960. *Metallanalysen Kupferzeitlicher und Frühbronzezeitlicher Bodenfunde aus Europa. Studien zu den Anfangen der Metallurgie*, Vol. 1. Mann Verlag, Berlin.

──────── 1968. *Kupfer und Bronze in der frühen Metallzeit Europas. Studien zu den Anfangen der Metallurgie*, Vol. 2. Mann Verlag, Berlin.

King, A. 1990. *Roman Gaul and Germany*. University of California Press, Berkeley.

Kingery, W. D. (ed.). 1986. *Technology and Style. Ceramics and Civilization: Ancient Technology to Modern Science*, Vol. 2. American Ceramic Society, Columbus, OH.

Krier, J., N. Theis, R. Wagner, and N. Folmer (eds.). 1986. *Carte archéologique du Grand-Duché de Luxembourg. Feuille 24: Differdange*. Musée d'Histoire et d'Art, Luxembourg.

Kruta, V. 1991. Celtic Writing. In *The Celts*, ed. S. Moscati, O. H. Frey, V. Kruta, B. Rafferty, and M. Szabó, pp. 491–498. Rizzoli, New York.

Kusimba, C. M., D. J. Killick, and R. G. Cresswell. 1994. Indigenous and Imported Metals at Swahili Sites on the Coast of Kenya. In *Society, Culture, and Technology in Africa*, ed. S. T. Childs, pp. 63–85. MASCA Research Papers in Science and Archaeology 11 (Supplement). MASCA, University of Pennsylvania Museum, Philadelphia.

Law, J. 1987. Technology and Heterogeneous Engineering: The Case of Portuguese Expansion. In *The Social Construction of Technological Systems*, ed. W. E. Bijker, T. P. Hughes and T. Pinch, pp. 111–134. MIT Press, Cambridge, MA.

Lechtman, H. 1977. Style in Technology—Some Early Thoughts. In *Material Culture: Style, Organization, and Dynamics*, ed. H. Lechtman and R. Merrill, pp. 3–20. West Publishing, St. Paul.

──────── 1979. Issues in Andean Metallurgy. In *Pre-Columbian Metallurgy of South America*, ed. E. P. Benson, pp. 1–40. Dumbarton Oaks, Washington, DC.

──────── 1980. The Central Andes: Metallurgy Without Iron. In *The Coming of the Age of Iron*, ed. T. A. Wertime and J. D. Muhly, pp. 267–334. Yale University Press, New Haven.

──────── 1984. Andean Value Systems and the Development of Prehistoric Metallurgy. *Technology and Culture* 25:1–36.

──────── 1988. Traditions and Styles in Central Andean Metalworking. In *The Beginning of the Use of Metals and Alloys*, ed. R. Maddin, pp. 344–378. MIT Press, Cambridge, MA.

──────── 1991. The Production of Copper-Arsenic Ores in the Central Andes: Highland Ores and Coastal Smelters? *Journal of Field Archaeology* 18:43–76.

Lechtman, H., and A. Steinberg. 1979. The History of Technology: An Anthropological Point of View. In *The History and Philosophy of Science*, ed. G. Bugliarello and D. B. Doner, pp. 135–162. University of Illinois Press, Urbana.

Lemonnier, P. 1986. The Study of Material Culture Today: Toward an Anthropology of Technical Systems. *Journal of Anthropological Archaeology* 5:147–186.

──────── 1989a. Towards an Anthropology of Technology. *Man*, n.s., 24(3):526–527.

──────── 1989b. Bark Capes, Arrowheads and Concorde: On Social Representations of Technology. In *The Meaning of Things: Material Culture and Symbolic Expression*, ed. I. Hodder, pp. 156–171. Unwin Hyman, London.

──────── 1990. Topsy Turvy Techniques: Remarks on the Social Representation of Techniques. *Archaeological Review from Cambridge* 9(1):27–37.

──────── 1992. *Elements for an Anthropology of Technology*. Museum of Anthropology, University of Michigan, Anthropological Paper 88. Ann Arbor.

Leroi-Gourhan, A. 1957. Le comportement technique chez l'animal et chez l'homme. In *L'evolution humaine*. Flammarion, Paris.

Levy, J. 1991. Metalworking Technology and Craft Specialization in Bronze Age Denmark. *Archaeomaterials* 5(1):55–74.

Liversage, D., and M. Liversage. 1989. A Method for the Study of the Composition of Early Copper and Bronze Artifacts: An Example from Denmark. *Helinium* 28:42–76.

Ludwig, R. 1988. Das frühromische Bandgräberfeld von Schankweiler, Kreis Bitburg-Prüm. *Trierer Zeitschrift* 51.51–422.

Manning, W. H. 1979. The Native and Roman Contribution to the Development of Metal Industries in Britain. In *Invasion and Response: The Case of Roman Britain*, ed. B. C. Burnham and H. B. Johnson, pp. 111–121. BAR British Series 73. British Archaeological Reports, Oxford.

Martin, P. S., G. I. Quimby, and D. Collier. 1947. *Indians Before Columbus*. University of Chicago Press, Chicago.

Mauss, M. 1935. Les techniques du corps. *Journal de Psychologie* 32:271–293.

McGovern, P. 1989. Ancient Ceramic Technology and Stylistic Change: Contrasting Studies from Southwest and Southeast Asia. In *Scientific Analysis in Archaeology*, ed. J. Henderson, pp. 63–81. UCLA Institute of Archaeology, Los Angeles.

Merrill, R. S. 1968. The Study of Technology. *International Encyclopedia of the Social Sciences*, Vol. 15, p. 577. New York.

Metzler, J. 1977. Beiträge zur Archäologie des Titelberges. In *Beiträge zur Archäologie und Numismatik des Titelberges*, Vol. 91, ed. J. Metzler and R. Weiller, pp. 17–103. Publication de la Section historique, Luxembourg.

_____ 1983. Fouilles du rempart de l'oppidum trévire du Titelberg. Les celts en Belgique et dans le nord de la France—Les fortifications de l'âge du fer. *Revue du Nord* special number.

_____ 1984. Das treverische Oppidum auf dem Titelberg (Luxemburg). In *Trier: Augustusstadt des Treverer*. Trier Stadtmuseum, Mainz.

_____ 1986. Le Titelberg: Oppidum trévire et vicus gallo-romain. In *Carte archéologique du Grand-Duché de Luxembourg*. Feuille 24: *Differdange*, ed. J. Krier, N. Theis, R. Wagner, and N. Folmer, pp. 22–29. Musée d'Histoire et d'Art, Luxembourg.

_____ 1991. The Celtic Horsemen's Graves at Goeblingen-Nospelt. In *The Celts*, ed. O. H. Frey, V. Kruta, B. Raftery, and M. Szabó, pp. 520–521. Rizzoli, New York.

Millett, M. 1990. Romanization: Historical Issues and Archaeological Interpretation. In *The Early Roman Empire in the West*, ed. T. Blagg and M. Millett, pp. 35–41. Oxbow Books, Oxford.

Montelius, O. 1903. *Die typologische Methode: Die älteren Kulturperioden im Orient und in Europa*, Vol. 1. Selbstverlag, Stockholm.

Nash, D. 1976. The Growth of Urban Society in France. In *Oppida: The Beginnings of Urbanisation in Barbarian Europe*, ed. B. Cunliffe and T. Rowley, pp. 95–133. BAR International Series 11. British Archaeological Reports, Oxford.

_____ 1978. Territory and State Formation in Central Gaul. In *Social Organisation and Settlement*, ed. D. Green, C. Haselgrove, and M. Spriggs, pp. 455–475. BAR International Series 47. British Archaeological Reports, Oxford.

_____ 1985. Celtic Territorial Expansion and the Mediterranean World. In *Settlement and Society: Aspects of Western European Prehistory in the First Millennium B.C.*, ed. T. C. Champion and J. V. S. Megaw, pp. 45–67. Leicester University Press, Leicester.

Neal, D. S, A. Wardle, and J. Hunn. 1990. *Excavation of the Iron Age, Roman and Medieval Settlement at Gorhambury, St. Albans*. English Heritage, London.

Nelson, M. C. 1991. The Study of Technological Organization. In *Archaeological Method and Theory*, ed. M. Schiffer, pp. 57–100. University of Arizona Press, Tucson.

Noble, D. F. 1984. *Forces of Production: A Social History of Industrial Automation*. Alfred A. Knopf, New York.

Northover, P. 1980. Bronze in the British Bronze Age. In *Aspects of Early Metallurgy*, ed. W. A. Oddy, pp. 63–70. British Museum Occasional Papers 7. London.

_____ 1985. The Complete Examination of Archaeological Metalwork. In *The Archaeologist and the Laboratory*, ed. P. Phillips, pp. 56–59. CBA Research Report 58. Council for British Archaeology, London.

_____ 1989. Non-Ferrous Metallurgy in Archaeology. In *Scientific Analysis in Archaeology*, ed. J. Henderson, pp. 213–236. UCLA Institute of Archaeology Research Tools 5. Los Angeles.

_____ 1992. Materials Issues in the Celtic Coinage. *Celtic Coinage: Britain and Beyond. 11th Oxford Symposium on Coinage and Monetary History*, ed. M. Mays, pp. 235–299. BAR British Series 222. British Archaeological Reports, Oxford.

Ortner, S. 1984. Theory in Anthropology Since the Sixties. *Comparative Studies in Society and History* 26(1):126–166.

Orton, C., P. Tyers, and A. Vince. 1993. *Pottery in Archaeology*. Cambridge University Press, Cambridge.

Oswalt, W. H. 1976. *An Anthropological Analysis of Food-Getting Technology*. John Wiley and Sons, New York.

Peacock, D. P. S. 1982. *Pottery in the Roman World: An Ethnoarchaeological Approach*. Longmans, London.

Pfaffenberger, B. 1988. Fetishised Objects and Humanised Nature: Towards an Anthropology of Technology. *Man*, n.s., 23:236–252.

_____ 1992. Social Anthropology of Technology. *Annual Review of Anthropology* 21:491–516.

Photos, E., S. J. Filippakis, and C. J. Salter. 1985. Preliminary Investigations of Some Metallurgical Remains at Knossos, Hellenistic to Third Century A.D. In *Furnaces and Smelting Technology in Antiquity*, ed. P.T. Craddock and M.J. Hughes, pp. 189–197. British Museum Occasional Papers 48. London.

Picon, M., J. Condamin, and S. Boucher. 1966.

Recherches techniques sur des bronzes de Gaule romaine. *Gallia* 24:189–214.

———— 1967. Recherches techniques sur des bronzes de Gaule romaine II. *Gallia* 25:153–168.

Piggott, S. 1965. *Ancient Europe from the Beginnings of Agriculture to Classical Antiquity*. Aldine, Chicago.

Pigott, V. C. 1980. The Iron Age in Western Iran. In *The Coming of the Age of Iron*, ed. T. A. Wertime and J. D. Muhly, pp. 417–461. Yale University Press, New Haven.

Pinch, T. J., and W. E. Bijker. 1987. The Social Construction of Facts and Artifacts: Or How the Sociology of Science and the Sociology of Technology Might Benefit Each Other. In *The Social Construction of Technological Systems*, ed. W. E. Bijker, T. P. Hughes, and T. Pinch, pp. 17–50. MIT Press, Cambridge, MA.

Pittioni, R. 1982. Twenty-Five Years of Spectroanalytical Research in Austria. *Journal of the Historical Metallurgy Group* 16(2):70–73.

Pocius, G. L. (ed.). 1991. *Living in a Material World: Canadian and American Approaches to Material Culture*. Social and Economic Papers 19. Institute of Social and Economic Research, Memorial University of Newfoundland. St. John's.

Pounds, N. J. 1973. *An Historical Geography of Europe, 450 B.C.–A.D. 1330*. Cambridge University Press, Cambridge.

Rabeison, E., and M. Menu. 1985. Métaux et alliages des bronziers d'Alesia. In *Recherches gallo-romaines* I, pp. 144–173. Laboratoire de Recherche des Musées de France, Paris.

Ralston, I. 1988. Central Gaul at the Roman Conquest: Conceptions and Misconceptions. *Antiquity* 62:786–794.

Raymond, R. 1986. *Out of the Fiery Furnace: The Impact of Metals on the History of Mankind*. Pennsylvania State University Press, University Park.

Reber, S. C., and M. R. Smith. 1986. Contextual Contrasts: Recent Trends in the History of Technology. In *High-Technology Ceramics: Past, Present, and Future*, ed. W. D. Kingery, pp. 1–15. American Ceramic Society, Columbus, OH.

Renfrew, C. 1986. Introduction: Peer Polity Interaction and Socio-Political Change. In *Peer Polity Interaction and Socio-Political Change*, ed. C. Renfrew and J. F. Cherry, pp. 1–18. Cambridge University Press, Cambridge.

Rice, P. 1981. The Evolution of Specialized Pottery Production: A Trial Model. *Current Anthropology* 22(3):219–240.

———— 1987. Economic Change in the Lowland Maya Late Classic Period. In *Specialization, Exchange and Complex Societies*, ed. E. M. Brumfiel and T. K. Earle, pp. 76–85. Cambridge University Press, Cambridge.

Riederer, J. 1974a. Metallanalysen römischer Sesterzen. *Jahrbuch für Numismatik und Geldgeschichte* 24:73–98.

———— 1974b. Römische Nähnadeln. *Technikgeschichte* 41:153–172.

———— 1983. Die Bedeutung der Metallanalyse für die Archäologie. In *Antidoren Jürgen Thimme*, ed. D. Metzler, B. Otto, and C. Müller-Wirth, pp. 159–164. C. F. Müller, Karlsruhe.

———— 1984. Metallanalysen römischen Bronzen. In *Toreutik und figürliche Bronzen römischer Zeit. Akten der 6. Tagung über antike Bronzen 13–17 Mai, Berlin*, pp. 220–225. Staatliche Museen, Preussischer Kulturbesitz, Berlin.

Riederer, J., and E. Briese. 1972. Metallanalysen römischer Gebrauchgegenstände. *Jahrbuch des Römisch-Germanischen Zentralmuseums* 19:83–88.

Riederer, J., and C. Laurenze. 1980. Metallanalysen römischer Henkel. *Berliner Beiträge Archäometrie* 5:37–42.

Romanoff, W., F. L. Ridell, and G. P. Halliwell. 1954. Copper-Base Foundry Alloys. In *Copper*, ed. A. Butts, pp. 508–534. Reinhold, New York.

Rostoker, W., and J. R. Dvorak. 1990. *Interpretation of Metallographic Structures*. Academic Press, San Diego.

Rothenberg, B., and A. B. Freijeiro. 1976. *The Huelva Archaeo-Metallurgical Project*. Institute for Archaeo-Metallurgical Studies, London.

———— 1980. Ancient Copper Mining and Smelting at Chinflon (Huelva, S.W. Spain). In *Scientific Studies in Early Mining and Extractive Metallurgy*, ed. P. T. Craddock, pp. 41–62. British Museum Occasional Papers 20. London.

Rousset, M., and M. Fedoroff. 1985. Multi-Element Instrumental NAA of Ag-Cu Coins. *Journal of Radio-Analytical and Nuclear Chemistry* 92(1):159–170.

Rowe, J. 1965. The Renaissance Foundation of Anthropology. *American Anthropologist* 67(1):1–20.

Rowlands, M. J. 1971. The Archaeological Interpretation of Prehistoric Metalworking. *World Archaeology* 3:210–224.

Rowlett, R. M., and A. L. Price. 1982. Differential Grain Use on the Titelberg, Luxembourg. *Journal of Ethnobiology* 2(1):79–88.

Rowlett, R. M., H. Thomas, and E.-J. Rowlett. 1982. Stratified Iron Age House Floors on the Titelberg, Luxembourg. *Journal of Field Archaeology* 9:301–312.

Roymans, N. 1990. *Tribal Societies in Gaul: An Anthropological Perspective*. Cingula 12. Universiteit

van Amsterdam, Amsterdam.
SSRC (Social Science Research Council). 1954. Acculturation: An Exploratory Formulation. *American Anthropologist*, n.s., 56:973–1002.
Sackett, J. R. 1982. Approaches to Style in Lithic Archaeology. *Journal of Anthropological Archaeology* 1:59–112.
Schiffer, M. 1992. *Technological Perspectives on Behavioral Change*. University of Arizona Press, Tucson.
Schiffer, M., and J. Skibo. 1987. Theory and Experiment in the Study of Technological Change. *Current Anthropology* 28(5):595–622.
Schlanger, N. 1990. Techniques as Human Action—Two Perspectives. *Archaeological Review from Cambridge* 9(1):18–26.
Scott, B. G. 1990. *Early Irish Ironworking*. Ulster Museum Publication 206. Belfast.
Scott, D. 1991. *Metallography and Microstructures of Ancient and Historic Metals*. J. Paul Getty Museum, Malibu.
Service, E. R. 1971. *Primitive Social Organization*. Random House, New York.
Shanks, M., and C. Tilley. 1987a. *Re-Constructing Archaeology*. Cambridge University Press, Cambridge.
—— 1987b. *Social Theory and Archaeology*. Polity Press, London.
Shepard, A. 1965. *Ceramics for the Archaeologist*. Carnegie Institute of Washington, Washington, DC.
Shrager, A. M. 1969. *Elementary Metallurgy and Metallography*, 3rd ed. Dover Publications, New York.
Singer, C. J. (ed.). 1954–58. *A History of Technology*. Clarendon Press, Oxford.
Slater, E. A., and J. A. Charles. 1970. Archaeological Classification by Metal Analysis. *Antiquity* 44:207–213.
Smith, C. S. 1967. The Interpretation of Microstructures of Metallic Artifacts. In *Application of Science in the Examination of Works of Art*, ed. W. J. Young, pp. 22–52. Museum of Fine Arts, Boston.
—— 1970. Discussion Session 1.1. In *Art and Technology, A Symposium on Classical Bronzes*, ed. S. Doeringer, D. G. Mitten, and A. Steinberg, pp. 51–56. MIT Press, Cambridge, MA.
—— 1981. *A Search for Structure: Essays on Science, Art, and History*. MIT Press, Cambridge, MA.
Speth, J. 1992. Foreword. In *Elements for an Anthropology of Technology*, by P. Lemonnier, pp. vii–ix. Museum of Anthropology, University of Michigan, Anthropological Paper 88. Ann Arbor.
Spicer, E. H. 1952. *Human Problems in Technological Change*. Russell Sage Foundation, New York.
Staudenmaier, J. M. 1985. *Technology's Storytellers: Reweaving the Human Fabric*. MIT Press, Cambridge, MA.
Steinberg, A. 1977. Technology and Culture: Technological Styles in the Bronzes of Shang China, Phrygia and Urnfield Central Europe. In *Material Culture: Styles, Organization, and Dynamics of Technology*, ed. H. Lechtman and R. S. Merrill, pp. 53–85. West Publishers, St. Paul, MN.
Steward, J. 1955. *Theory of Culture Change*. University of Illinois Press, Urbana.
Stewart, E. M. 1950. *Vounous 1937–38: Field Report on the Excavations Sponsored by the British School of Archaeology at Athens*. CWK Gleerup, Lund.
Stos-Gale, Z. 1989. Lead-Isotope Studies of Metals and the Metal Trade in the Bronze Age Mediterranean. In *Scientific Analysis in Archaeology*, ed. J. Henderson, pp. 274–301. UCLA Institute of Archaeology, Los Angeles.
Stutzinger, D. 1984. Die Sammlung Antiker Bronzegefässe des Römisch-Germanischen Museums in Köln—Materialanalysen and Archäologische Daten. In *Toreutik und Figürliche Bronzen Römischer Zeit: Akten der 6. Tagung über Antike Bronzen 13–17 Mai in Berlin*, ed. Staatliche Museen, pp. 232–238. Preussischer Kulturbesitz, Berlin.
Swann, C., and S. Fleming. 1986. New Directions in the Bartol PIXE System for Studies in Archaeometry. *Nuclear Instruments and Methods in Physics Research* 14:61–69.
Tacitus, C. 1971. *The Agricola and the Germania*, transl. H. Mattingly. Penguin Books, Harmondsworth.
Tchernia, A. 1983. Italian Wine in Gaul at the End of the Republic. In *Trade in the Ancient Economy*, ed. P. Garnsey, K. Hopkins, and C. R. Whittaker, pp. 87–104. University of California Press, Berkeley.
Thill, G. 1965. *Tetelberg: Site archéologique*. Musée d'Histoire et d'Art, Luxembourg.
—— 1967. Die Keramik aus vier spätlatènezeitlichen Brandgräbern von Goeblingen-Nospelt. *Hèmecht* 19:204–209.
Thomas, H. L., and R. M. Rowlett. 1979. Preliminary Results of the Titelberg Project. In *The 1978–9 Annual Report of the Museum of Anthropology, University of Missouri-Columbia*, ed. L. H. Feldman and E. S.-J. Rowlett, pp. 50–71. Museum of Anthropology, Columbia, MO.
Thomas, H. L., R. M. Rowlett, and E. S.-J. Rowlett. 1975. The Titelberg: A Hill Fort of Celtic and Roman Times. *Archaeology* 28(1):55–57.
—— 1976. Excavations on the Titelberg, Luxembourg. *Journal of Field Archaeology* 3:241–259.

Tierney, J. 1960. The Celtic Iconography of Posidonius. *Proceedings of the Royal Irish Academy* 60:189–275.

Tilley, C. 1984. Ideology and the Legitimation of Power in the Middle Neolithic of Southern Sweden. In *Ideology, Power and Prehistory*, ed. D. Miller and C. Tilley, pp. 111–146. Cambridge University Press, Cambridge.

———— 1990. *Reading Material Culture: Structuralism, Hermaneutics and Post-Structuralism*. Basil Blackwell, Oxford.

Torrence, R. (ed.). 1989. *Time, Energy and Stone Tools*. Cambridge University Press, Cambridge.

Tournaire, J., O. Buchsenshutz, J. Henderson, and J. Collis. 1982. Iron Age Coin Moulds from France. *Proceedings of the Prehistoric Society* 48:417–435.

Trigger, B. 1986. The Role of Technology in V. Gordon Childe's Archaeology. *Norwegian Archaeological Review* 19(1):1–14.

———— 1989. *A History of Archaeological Thought*. Cambridge University Press, Cambridge.

Tylecote, R. F. 1970. The Composition of Metal Artifacts: A Guide to Provenance? *Antiquity* 44:19–25.

———— 1976. *A History of Metallurgy*. The Metals Society, London.

———— 1985. The Apparent Tinning of Bronze Axes and Other Artefacts. *Journal of the Historical Metallurgy Society* 19(2):169–175.

———— 1986. *The Prehistory of Metallurgy in the British Isles*. Institute of Metals, London.

———— 1987. *The Early History of Metallurgy in Europe*. Longman, London.

Tylecote, R. F., H. A. Ghaznavi, and R. J. Boydell. 1977. Partitioning of Trace Elements between Ores, Fluxes, Slags and Metals during the Smelting of Copper. *Journal of Archaeological Sciences* 4(4):305–333.

Van der Leeuw, S. E. 1983. Acculturation as Information Processing. In *Roman and Native in the Low Countries: Spheres of Interaction*, ed. R. Brandt and J. Slofstra, pp. 11–42. BAR International Series 184. British Archaeological Reports, Oxford.

Vandiver, P., and G. S. Wheeler. 1991. Introduction to Materials Issues of Art and Archaeology. In *Materials Issues in Art and Archaeology*, Vol. 2, ed. P. Vandiver, J. Druzik, and G. Wheeler. Symposium Proceedings of the Materials Research Society, vol. 185. Pittsburgh.

Wade, J. A. 1989. The Context of Adoption of Brass Technology in Northeastern Nigeria and Its Effects on the Elaboration of Culture. In *What's New: A Closer Look at the Process of Innovation*, ed. S. E. van der Leeuw and R. Torrence, pp. 225–244. Unwin Hyman, London.

Wainwright, G., and M. Spratling. 1973. The Iron Age Settlement of Gussage All Saints. *Antiquity* 47:109–129.

Wallace, A. F. C. 1972. Paradigmatic Processes in Culture Change. *American Anthropologist* 74:467–478.

Waterbolk, H. T., and J. J. Butler. 1965. Comments on the Use of Metallurgical Analysis in Prehistoric Studies. *Helinium* 5:227–251.

Weaver, V. P. 1954. Wrought Copper Alloys. In *Copper*, ed. A. Butts, pp. 535–572. Reinhold, New York.

Weiller, R. 1977. Die Münzfunde aus der Grabung vom Tetelbierg. In *Beiträge zur Archäologie und Numismatik des Titelberges*. Vol. 91, ed. J. Metzler and R. Weiller, pp. 118–195. Publications de la Section Historique, Luxembourg.

———— 1979. Les techniques de fabrication employées dans l'atelier monétaire de l'oppidum trévire du Tetelbierg. *Actes du 9ème Congrès International de Numismatique, Berne*, pp. 625–632. AINP Publication 6, Luxembourg.

———— 1986. L'atelier monètaire trévire du Tetelbierg. *Carte archéologique du Grand-Duché de Luxembourg*. Feuille 24: *Differdange*, ed. J. Krier, N. Theis, R. Wagner, and N. Folmer, pp. 30–33. Musée d'Histoire et d'Art, Luxembourg.

Weiner, A., and J. Schneider (eds.). 1989. *Cloth and Human Experience*. Smithsonian Institution Press, Washington, DC.

Wells, P. 1984. *Farms, Villages and Cities: Commerce and Urban Origins in Late Prehistoric Europe*. Cornell University Press, Ithaca.

White, L. A. 1949. *The Science of Culture*. Farrar Straus and Cudahy, New York.

White, L. 1962. *Medieval Technology and Social Change*. Clarendon Press, Oxford.

Wightman, E. M. 1971. *Roman Trier and the Treveri*. Praeger, New York.

———— 1985. *Gallia Belgica*. University of California Press, Berkeley.

Wild, J. P. 1976. Loanwords and Roman Expansion in North-West Europe. *World Archaeology* 8(1):57–64.

Woimant, G.-P., and L. P. Hurtel. 1989. Statuette de dieu-guerrier gaulois. In *Les bronzes antiques de Paris*, ed. J. Bonnet, P. de Carbonniere, L. Faudin, P. Forni, G. Garriga, N. Morand-van Haecke, and P. Velay, pp. 465–468. Musée Carnavalet, Paris.

Wright, R. A. 1985. Technology and Style in Ancient Ceramics. In *Ceramics and Civilization: Ancient Technology to Modern Science*, ed. W. D. Kingery, pp. 5–26. American Ceramic Society, Columbus, OH.

———— 1986. The Boundaries of Technology and Stylistic Change. In *Technology and Style. Ceramics and Civilization: Ancient Technology to Modern Science*, Vol. 2, ed. W. D. Kingery, pp. 1–20.

American Ceramic Society, Columbus, OH.

Yener, K. A., E. V. Sayer, E. C. Joel, H. Özbal, I. C. Barnes, and R. H. Brill. 1991. Stable Lead Isotope Studies of Central Taurus Ore Sources and Related Artifacts from Eastern Mediterranean Chalcolithic and Bronze Age Sites. *Journal of Archaeological Sciences* 18(5):541–577.

Young, R. 1981. *The Gordion Excavations, Final Reports*. Vol. 1: *Three Great Early Tumuli*. University of Pennsylvania Museum Monograph 43. Philadelphia.

Zwicker, U., H. Greiner, K.-H. Hoffmann, and M. Reithinger. 1985. Smelting, Refining and Alloying of Copper and Copper Alloys in Crucible-Furnaces During Prehistoric up to Roman Times. In *Furnaces and Smelting Technology in Antiquity*, ed. P. T. Craddock and M. J. Hughes, pp. 103–115. British Museum Occasional Papers 48. London.

APPENDIX

COMPOSITIONAL DATA
FROM ANALYZED ARTIFACTS
FROM THE TITELBERG

APPENDIX
COMPOSITIONAL DATA FOR ANALYZED ARTIFACTS FROM THE TITELBERG

ID #	Period	Artifact	Cu	Sn	Pb	Zn	Fe	As	Ag	Sb	S	Ni	Cl
0421-82	1	debris	81.6	16.3	0.124	≤0.39	0.116	0.058	0.474	0.259	0.206	0.119	0.040
0455-82	1	debris	84.0	11.3	1.35	≤0.44	0.074	0.127	0.127	0.161	0.564	0.124	0.681
0988-74a	1	shaft	88.4	6.7	2.89	≤0.39	0.059	0.148	0.133	0.232	0.528	0.214	0.024
0988-74b	1	fitting	85.7	12.7	0.078	≤0.52	0.139	0.102	0.060	0.188	0.029	0.151	0.013
0125-74	2	tool	87.3	9.7	≤0.034	≤0.38	0.289	0.023	0.067	0.219	0.100	0.201	0.215
0226-74	2	fitting	90.0	8.8	≤0.057	≤0.38	0.139	0.021	0.090	0.060	0.031	0.126	0.009
0372-74	2	shaft	85.6	12.8	0.095	≤0.46	0.056	0.029	0.193	0.212	0.051	0.183	0.019
0431-82	2	debris	87.1	10.5	0.606	≤0.32	0.060	0.173	0.150	0.193	0.251	0.229	0.038
0469-72	2	tack/rivet	78.0	5.8	11.7	≤0.43	0.325	0.351	0.454	0.799	0.060	0.178	0.085
0504-82	2	fibula	82.8	2.1	0.278	13.6	0.240	0.055	0.079	≤0.020	0.272	0.097	0.110
0703-73	2	fitting	80.4	0.31	0.474	18.2	0.229	≤0.016	0.058	≤0.014	0.107	0.084	0.025
0720-73	2	fibula	77.3	0.020	≤0.047	21.9	0.094	≤0.008	0.054	≤0.044	0.066	0.085	0.053
1053-73	2	fitting	78.6	0.28	0.573	19.9	0.164	0.035	0.041	≤0.020	0.135	0.085	0.149
0010-78	3	tack/rivet	97.5	0.31	≤0.060	≤0.37	0.067	≤0.017	0.386	0.361	0.032	0.498	0.188
0043-82	3	debris	80.9	14.7	0.076	≤0.36	0.069	0.023	0.655	0.658	1.22	0.090	0.632
0063-78	3	fitting	98.9	0.020	≤0.022	≤0.40	0.046	≤0.007	≤0.010	≤0.014	0.135	0.078	0.131
0107-74	3	tack/rivet	98.1	0.021	≤0.026	≤0.44	0.078	≤0.007	0.283	0.070	0.224	0.497	0.050
0125-82	3	debris	79.0	19.2	0.089	≤0.37	0.042	0.092	0.263	0.183	0.142	0.120	0.207
0188-77	3	debris	82.0	15.2	0.356	≤0.37	0.092	0.067	0.176	0.369	0.463	0.170	0.364
0192-78	3	fitting	65.7	6.3	25.7	≤0.41	0.056	0.684	0.093	0.198	≤0.015	0.125	0.047
0269-82	3	fitting	80.3	0.73	0.169	18.0	0.154	≤0.012	0.070	0.074	0.101	0.142	0.025
0287-82	3	debris	87.0	5.4	5.59	≤0.38	0.066	0.203	0.219	0.347	≤0.007	0.204	0.016
0319-76	3	tack/rivet	87.1	11.1	0.659	≤0.40	0.051	0.061	0.216	≤0.042	0.058	0.102	0.006
0338-78	3	fibula	79.4	1.7	0.959	17.0	0.331	0.030	5.04	≤0.020	≤0.004	0.062	0.097
0352-82a	3	debris	82.9	13.7	0.135	≤0.36	0.083	0.503	0.411	0.173	0.282	0.131	0.982
0389-73	3	fitting	82.1	0.22	0.234	16.7	0.120	≤0.014	0.067	≤0.022	0.062	0.109	0.045
0394-78	3	fitting	91.1	7.2	0.105	≤0.38	0.077	≤0.011	0.038	0.063	0.206	0.085	0.078
0430-82	3	debris	85.0	11.5	1.36	≤0.45	0.038	0.129	0.128	0.163	0.331	0.126	0.575
0482-78	3	fibula	97.8	0.033	0.180	≤0.39	0.052	≤0.010	0.821	0.439	0.032	0.126	0.006
0483-82	3	shaft	85.9	11.1	0.263	≤0.37	0.203	0.058	0.107	0.128	0.369	0.091	0.516
0496-72	3	fibula	78.6	0.82	0.196	18.9	0.199	0.025	0.082	0.051	0.361	0.151	0.217
0503-78	3	fitting	75.2	13.5	8.52	≤0.34	0.043	0.300	0.190	0.714	0.043	0.131	0.281
0508-73a.1	3	tool	76.9	0.41	0.074	21.9	0.126	0.040	0.032	≤0.017	0.136	0.087	0.151
0508-73b	3	fitting	91.5	6.5	0.694	≤0.40	0.108	0.110	0.071	≤0.044	0.153	0.157	0.027
0509-78a.1	3	fibula	85.4	1.5	0.165	11.5	0.166	0.032	0.070	0.128	0.193	0.091	0.158

ID #	Period	Artifact	Cu	Sn	Pb	Zn	Fe	As	Ag	Sb	S	Ni	Cl
0509-78b	3	tack/rivet	76.1	4.0	4.21	14.1	0.200	0.198	0.102	0.218	0.044	0.107	0.217
0521-78	3	fibula	78.3	0.53	0.365	19.8	0.281	0.020	≤0.014	≤0.017	0.191	0.075	0.130
0523-78	3	tool	98.3	0.28	≤0.028	≤0.39	0.079	0.083	0.037	0.422	0.039	0.106	0.101
0525-78	3	tool	79.0	0.92	0.330	19.0	0.161	≤0.014	0.205	≤0.020	0.104	0.130	0.020
0528-72	3	fitting	67.6	12.7	10.3	≤0.33	0.048	0.973	0.843	6.24	0.044	0.391	0.038
0724-73	3	shaft	89.8	6.4	0.662	≤0.35	0.107	0.122	0.089	0.051	0.285	0.150	0.050
0727-77	3	tack/rivet	87.4	11.3	0.265	≤0.39	0.064	0.055	0.118	0.099	0.031	0.151	0.007
0727-78b	3	tack/rivet	98.0	0.012	≤0.029	≤0.37	0.058	0.554	≤0.009	≤0.017	0.150	0.460	0.024
0753-77	3	fitting	88.6	9.9	0.294	≤0.38	0.059	0.047	0.135	0.144	0.040	0.179	0.006
0790-77	3	fitting	79.5	0.51	0.325	19.1	0.192	≤0.012	0.075	≤0.016	0.102	0.084	0.013
0809-78	3	tack/rivet	79.1	0.99	0.295	18.8	0.209	0.100	0.060	≤0.048	0.082	0.179	0.021
0875-73a	3	fibula	74.1	0.35	≤0.055	22.7	0.120	0.123	0.009	≤0.027	0.636	0.084	0.724
0907-73	3	fibula	79.2	6.8	9.98	≤0.34	0.134	0.376	0.913	1.38	≤0.009	0.270	0.019
0913-73	3	fibula	77.1	0.30	0.110	21.6	0.128	≤0.016	0.050	≤0.014	0.035	0.271	0.140
0962-73	3	fibula	93.4	5.5	0.108	≤0.47	0.093	0.045	0.085	0.071	0.020	0.145	0.005
0984-77	3	tack/rivet	97.7	0.015	≤0.022	≤0.35	0.066	1.11	0.084	0.238	0.016	0.242	0.007
1013-73a	3	fitting	85.0	11.7	0.229	0.51	0.361	0.072	0.122	0.091	0.811	0.143	0.147
1013-73b	3	fitting	84.5	13.7	0.266	≤0.58	0.182	0.080	0.077	0.077	0.048	0.141	0.013
1013-73c	3	fitting	85.8	11.3	0.194	≤0.40	0.211	0.063	0.090	0.111	1.15	0.146	0.098
1083-77	3	tack/rivet	98.2	0.019	≤0.021	≤0.35	0.074	≤0.006	0.260	0.073	≤0.019	0.426	0.025
1249-77	3	shaft	87.7	7.5	1.43	≤0.46	0.231	0.263	0.045	0.034	0.056	0.134	0.531
0004-78	4	fitting	82.3	0.042	0.134	16.9	0.255	0.027	0.069	≤0.011	0.081	0.092	0.018
0031-81	4	fitting	65.8	22.4	9.75	≤0.59	0.104	0.160	0.214	0.376	0.022	0.109	0.055
0069-77	4	fitting	79.1	0.27	0.185	19.7	0.309	≤0.015	≤0.015	≤0.014	0.045	0.068	0.028
0071-72	4	fibula	80.4	0.87	0.255	18.0	0.147	≤0.012	0.080	≤0.021	≤0.003	0.089	0.009
0137-77	4	fitting	98.9	0.022	≤0.028	≤0.37	0.043	0.041	0.066	≤0.017	0.215	0.131	0.036
0159-77	4	fitting	75.8	0.60	0.238	22.8	0.075	0.029	0.116	≤0.022	0.025	0.086	0.013
0176-73	4	fibula	74.7	0.074	0.325	24.3	0.164	0.044	0.059	≤0.049	0.040	0.085	0.008
0207-72	4	fibula	73.7	0.029	≤0.068	25.0	0.248	0.022	0.037	0.086	0.043	0.082	0.058
0214-82	4	fibula	76.2	0.20	≤0.064	22.0	0.219	0.032	0.048	≤0.044	0.096	0.092	0.024
0223-73	4	tack/rivet	94.9	0.013	0.151	≤0.39	0.241	≤0.011	0.049	0.058	1.11	0.094	0.947
0273-72	4	fitting	75.8	2.3	6.36	14.5	0.177	0.106	0.031	≤0.028	0.025	0.083	0.277
0289-73	4	fibula	77.0	1.2	≤0.028	20.5	0.254	0.031	0.033	≤0.038	0.189	0.072	0.132
0309-77	4	shaft	97.0	0.014	≤0.032	≤0.38	0.131	0.115	0.039	0.454	0.422	0.090	0.322
0321-72	4	fibula	78.4	0.244	0.775	19.6	0.202	0.036	0.087	0.094	0.185	0.154	0.009
0323-72	4	fitting	80.9	0.45	0.212	17.7	0.317	0.026	0.060	≤0.020	0.086	0.070	0.020
0346-82	4	fitting	68.4	23.1	5.64	≤0.31	0.047	0.390	0.322	1.01	0.022	0.153	0.035
0384-73	4	fibula	75.0	0.42	0.089	21.6	0.131	0.121	≤0.010	≤0.026	0.658	0.087	0.798

ID #	Period	Artifact	Cu	Sn	Pb	Zn	Fe	As	Ag	Sb	S	Ni	Cl
0397-72	4	fibula	80.7	4.0	1.82	12.4	0.217	0.123	0.076	0.094	0.241	0.099	0.010
0400-73	4	fibula	81.2	15.1	0.494	≤0.38	0.120	0.147	0.374	0.196	0.755	0.122	0.669
0505-77	4	tack/rivet	75.0	0.32	≤0.035	23.9	0.100	0.037	≤0.013	≤0.022	0.026	0.201	0.131
0583-77	4	shaft	92.7	6.1	0.092	≤0.42	0.120	0.060	0.053	0.050	0.023	0.135	0.005
0584-77	4	fitting	77.9	0.028	≤0.055	20.7	0.181	≤0.008	0.060	0.054	0.038	0.081	0.050
0605-74a	4	tack/rivet	96.7	0.029	≤0.026	≤0.42	0.064	≤0.010	0.575	0.514	0.165	0.501	0.134
0605-74b	4	tack/rivet	97.6	0.012	≤0.028	≤0.38	0.414	0.156	0.035	≤0.016	0.582	0.121	0.113
0606-74	4	tack/rivet	92.8	3.4	≤0.029	≤0.35	0.150	0.321	0.080	0.061	0.032	0.658	0.052
0638-73	4	tack/rivet	72.7	5.7	13.4	6.84	0.171	0.170	0.066	0.075	0.052	0.090	0.039
0678-78	4	debris	86.5	11.2	0.372	≤0.32	0.043	0.065	0.178	0.230	0.275	0.146	0.145
0708-77a	4	fibula	76.6	1.30	0.078	21.4	0.152	0.031	0.066	0.139	0.027	0.096	0.003
0708-77b	4	tack/rivet	69.7	6.2	7.44	14.3	0.379	0.398	0.097	0.130	≤0.011	0.118	0.247
0780-73	4	fibula	81.9	12.2	1.83	≤0.41	0.053	0.196	0.407	2.41	0.203	0.084	0.007
0813-77	4	fitting	78.7	0.092	≤0.051	17.4	0.263	≤0.011	0.034	0.067	0.455	0.080	0.546
0880-77	4	fitting	85.0	12.2	0.688	≤0.36	0.060	0.131	0.783	0.180	0.114	0.129	0.094
0901-77	4	fitting	67.4	5.4	25.5	≤0.46	0.054	0.291	0.146	≤0.046	≤0.014	0.083	0.053
0905-77	4	tack/rivet	86.1	10.8	1.15	≤0.36	0.078	0.122	0.121	0.436	0.174	0.167	0.082
0912-77b	4	fibula	81.3	4.0	0.116	13.6	0.135	0.031	0.061	0.055	0.173	0.090	0.216
0918-77	4	tack/rivet	83.4	1.1	0.127	12.3	0.299	0.059	0.120	0.131	0.164	0.079	0.147
0967-77	4	debris	82.6	11.6	2.27	≤0.37	0.109	0.300	0.170	0.297	0.600	0.159	0.319
0987-77	4	fibula	77.2	0.14	0.092	21.4	0.175	≤0.016	0.054	≤0.040	0.062	0.080	0.037
0990-77	4	fibula	77.4	0.53	≤0.056	21.0	0.140	≤0.019	0.045	0.142	0.059	0.091	0.028
0998-77	4	shaft	86.5	10.6	≤0.061	≤0.37	0.351	≤0.012	0.056	≤0.038	0.159	0.072	0.080
0999-77	4	fibula	79.7	0.54	≤0.054	18.8	0.185	0.182	0.044	0.055	0.130	0.088	0.048
1031-73	4	shaft	97.3	0.029	0.341	≤0.48	0.049	≤0.015	0.557	0.406	0.325	0.215	0.531
1043-77	4	tack/rivet	75.2	0.31	5.42	21.3	0.206	0.026	0.042	0.052	0.771	0.082	0.976
0009-81	5	shaft	86.6	0.17	0.257	11.8	0.276	0.042	0.059	0.132	0.200	0.102	0.137
0113-73	5	fibula	74.6	0.51	0.158	22.5	0.294	0.029	0.041	0.104	0.340	0.086	0.239
0133-73	5	fitting	71.1	0.71	10.3	16.5	0.203	0.549	≤0.020	0.060	≤0.007	0.115	0.046
0182-72	5	fibula	76.3	0.64	0.503	20.3	0.276	0.106	0.065	≤0.028	0.185	0.111	0.086
0229-74b	5	tack/rivet	96.8	0.017	0.343	≤0.43	0.153	0.021	0.153	0.073	0.362	0.166	0.479
0318-77	5	shaft	77.5	7.5	8.82	4.56	0.291	0.125	0.335	0.090	0.027	0.091	0.084
0360-73	5	fitting	96.8	2.0	≤0.047	≤0.49	0.132	0.027	0.053	0.080	0.056	0.101	0.009
0395-74i	5	fitting	84.7	3.1	0.487	10.8	0.212	0.068	0.073	0.158	0.089	0.110	0.011
0395-74o	5	tack/rivet	96.4	1.6	0.339	≤0.57	0.183	0.039	0.125	0.079	0.054	0.101	0.044
0441-74	5	fitting	80.9	0.37	0.494	17.4	0.325	0.041	0.066	≤0.017	0.102	0.075	0.010
0446-76	5	fibula	82.4	2.5	0.534	13.5	0.377	0.055	0.041	0.077	0.162	0.094	0.014
0487-76	5	debris	80.2	9.3	5.89	2.86	0.129	0.367	0.225	0.156	≤0.007	0.119	0.027

ID #	Period	Artifact	Cu	Sn	Pb	Zn	Fe	As	Ag	Sb	S	Ni	Cl
0512-76	5	fibula	83.2	2.1	2.50	11.2	0.233	0.166	0.067	0.215	≤0.005	0.111	0.011
0525-76	5	fitting	62.0	4.3	24.5	7.77	0.421	0.229	0.070	≤0.042	0.061	0.074	0.048
0661-76	5	fibula	78.5	1.5	0.087	19.2	0.137	0.045	0.062	0.070	0.049	0.098	0.020
0730-76a	5	fibula	84.4	0.010	0.093	15.0	0.142	≤0.018	≤0.026	≤0.038	0.047	0.106	0.019
0730-76b	5	shaft	90.2	8.4	0.291	≤0.37	0.082	0.190	0.054	≤0.031	0.069	0.114	0.007
0732-76	5	fitting	81.3	0.069	≤0.038	18.1	0.079	≤0.012	0.054	0.065	0.031	0.072	0.026
0754-76	5	tool	79.2	0.35	0.139	19.0	0.349	≤0.011	0.056	≤0.016	0.133	0.083	0.048
0794-76	5	fibula	77.6	0.26	0.100	21.3	0.234	0.052	0.060	0.064	0.037	0.075	0.013
0796-77	5	fibula	82.3	0.83	0.349	15.1	0.319	0.055	0.044	≤0.034	0.078	0.084	0.026
0853-73	5	fitting	76.3	0.24	≤0.052	20.2	0.217	0.034	0.075	≤0.017	0.470	0.096	0.676

MASCA Research Papers in Science and Archaeology

Volume 13, 1996

Series Editor
Kathleen Ryan

Production Editors
Helen Schenck
Jennifer Quick

Advisory Committee
Stuart Fleming, Chairman
Philip Chase
Patrick McGovern
Henry Michael
Naomi F. Miller
Vincent Pigott

Design and Layout
Helen Schenck

Graphics
Paul Zimmerman

**Customer Service/
Subscription Manager**
Tony DeAnnuntis

The subscription price for *MASCA Research Papers in Science and Archaeology* is $20, payable in U.S. dollars. We also accept VISA/MASTERCARD. This price covers one main volume per year. In addition, we publish supplementary volumes, which are offered to MASCA subscribers at a discounted price.

This is a refereed series. All materials for publication should be sent to The Editor, *MASCA Research Papers in Science and Archaeology*. Subscription correspondence should be addressed to The Subscription Manager, MASCA, University of Pennsylvania Museum, 33rd and Spruce Streets, Philadelphia, PA 19104-6324.